Sustaining
Tomorrow

Sustaining
Tomorrow

A STRATEGY FOR WORLD CONSERVATION AND DEVELOPMENT

Edited by
Francis R. Thibodeau
Hermann H. Field

Published for Tufts University by
University Press of New England
HANOVER AND LONDON

University Press of New England

BRANDEIS UNIVERSITY
BROWN UNIVERSITY
CLARK UNIVERSITY
DARTMOUTH COLLEGE
UNIVERSITY OF NEW HAMPSHIRE
UNIVERSITY OF RHODE ISLAND
TUFTS UNIVERSITY
UNIVERSITY OF VERMONT

Printed in the United States of America

LIBRARY OF CONGRESS CATALOGING
IN PUBLICATION DATA
Main entry under title:

Sustaining tomorrow.

Bibliography: p. 172.
Includes index.
1. Environmental policy—Addresses, essays,
lectures.   2. Conservation of natural resources—
Addresses, essays, lectures.   3. Economic
development—Environmental aspects —Addresses,
essays, lectures.   I. Thibodeau, Francis R., 1953-
         II. Field, Hermann H., 1910-
III. Tufts University.
HC79.E5s87   1984   333.7'2   84-40297
ISBN 0-87451-305-7
ISBN 0-87451-306-5 (pbk.)

Contents

Contents

Foreword

It gives me great pleasure to introduce *Sustaining Tomorrow*. It is genuinely difficult to imagine a more distinguished group of experts addressing a more vital set of concerns. They have taken on no less a task than examining the care of an entire planet and all its inhabitants.

In March of 1980 the International Union for Conservation of Nature and Natural Resources launched the World Conservation Strategy in co-operation with the U.N. Environment Programme and the World Wildlife Fund. In all, nearly 1,000 scientists and other experts helped to create this remarkable document. It differs in three ways from anything that has ever been attempted before.

It examines the full range of challenges facing us as stewards of a fragile planet.

It points the way toward concrete solutions to these challenges. It is a true strategy, a program for action.

Perhaps most important, it stresses the unity of resource conservation and human development as mutually reinforcing tools for building a sustainable future.

This last point may need some explanation. In the past conservation and development have been seen too often as opposing forces—either you were "for" wildlife and wild places or you were "for" human development. But it is not difficult to see that the same problems that threaten the biosphere threaten all of us. When the last of a wild species is destroyed, we should not only feel the loss of an irreplaceable aspect of the natural world, we should also feel the loss of the potential it may have had to provide new medicines, new crops, new products for a better human existence. Similarly, when the tropical forests are wantonly destroyed, we not only lose a great treasure of biological diversity, we also lose the stabilizing effect those forests have on our climate. Conservation and development are not only compatible, they are two expressions of the same need—to keep the earth as a sustaining home.

The strategy received immediate support from governments of nearly every political persuasion—China, India, Spain, Jordan, Egypt, Kenya, and Australia among them—the European Parliament at Strasbourg, and the European Economic Community in Brussels. Official support is still growing. Many parts of the world have already produced their own regional conservation strategies as policymakers, their advisers, conservationists, development agencies, industries, and unions realize the impact the strategy is having on the resolution of environmental problems. But there are still important groups who must have the opportunity to hear about the World Conservation Strategy: conservationists, development professionals, government officials, those involved in education and training, and concerned individuals who work at the local or regional level.

This book is intended to bring the message of the World Conservation Strategy to them. In a very real sense, our success in meeting the goals set forth in the strategy will depend on their ability to make it a foundation for local action. This book is also intended for the students and teachers to whom we will pass on the charge of global stewardship. They will find much left to be done. We hope this volume can serve as a resource book for both in-service and public education.

Each of the contributors to this book is an internationally recognized expert in some aspect of environmental science or policy. All present the most recent information and offer exciting, often surprising, interpretations of the responses these data demand. Together, they cover the full range of problems—and opportunities—we face as we assume greater responsibility for managing the biosphere.

This book is an invitation to those who will take on that responsibility, especially those working in their own localities. But even the best books can only invite, inform, and guide. The challenge of integrating conservation and development still rests where it always has, with those who will work toward building a more satisfying and sustainable relationship between society and the environment.

HAROLD J. COOLIDGE
Honorary President
International Union for
Conservation of Nature
and Natural Resources

Preface

In the early 1970s environmental conservation changed. Work that had been carried on by a relatively small number of dedicated experts since World War II suddenly took hold in the imagination of the general public. Both the media and governments began to pay increased attention to environmental problems. Perhaps more important, many people who had never before been active in environmental affairs started to press for change within their own communities. The efforts of a few dedicated professionals suddenly expanded into a grass roots movement.

During the late 1970s this new power and visibility brought many positive results. But it also brought problems. Each of the many new coalitions set its priorities and organized support in relative isolation. The very strength of these groups' focus on issues within a familiar home territory gave them little reason to consider the broad implications of their collective actions. Moreover, among those who had not become convinced of the urgent necessity of expanded environmental action, there was a growing suspicion that conservationists had forgotten the importance of meeting human needs. In some quarters conservationists began to be viewed as disorganized obstructionists.

Although this assessment was strikingly inaccurate in all but a very few cases, it was obvious that the conservation community needed to develop a broad statement of purpose, not only to answer critics, but also to set a clear internal agenda. Any such statement would have to serve three functions: first, it would have to explain the nature and magnitude of the threat posed by present unsound resource policies; second, it would have to demonstrate the relevance of conservation to the prob-

Portions of the preface have appeared in the *World Conservation Strategy* copyright 1980 IUCN-UNEP-WWF and "The World Conservation Strategy and Local Environmental Action" Environmental Education and Information copyright 1982 Taylor and Francis, Ltd.

lems impeding development; and third, it would have to present a clear plan of action.

In 1980 the International Union for Conservation of Nature and Natural Resources (IUCN), together with the United Nations Environment Programme (UNEP) and the World Wildlife Fund (WWF), produced the needed document. *World Conservation Strategy: Living Resources for Sustainable Development* heralds the newest phase in the development of the conservation movement. It signals a reunification among conservationists, no matter what the geographic scale of their action, and it clearly demonstrates that conservation and development must be inseparable parts of a single effort to build a sustainable future.

The strategy was initiated when IUCN asked 450 governmental and nongovernmental agencies in more than 100 countries to define their environmental concerns. Drafts of a document based on the replies were sent to more than 700 experts on ecology, endangered species, natural areas, environmental education, environmental law, and environmental planning. After two years of synthesis and review, the strategy was made public on 5 March 1980. The immediate response was extremely enthusiastic.

The executive summary of the World Conservation Strategy is perhaps the best preface both to this book and to contemporary conservation thinking:

1. The aim of the World Conservation Strategy is to achieve the three main objectives of living resource conservation:

a. to maintain essential ecological processes and life-support systems (such as soil regeneration and protection, the recycling of nutrients, and the cleansing of waters), on which human survival and development depend;

b. to preserve genetic diversity (the range of genetic material found in the world's organisms), on which depend the functioning of many of the above processes and life-support systems, the breeding programmes necessary for the protection and improvement of cultivated plants, domesticated animals and microorganisms, as well as much scientific and medical advance, technical innovation,

and the security of the many industries that use living resources;

c. to ensure the sustainable utilization of species and ecosystems (notably fish and other wildlife, forests and grazing lands), which support millions of rural communities as well as major industries.

2. These objectives must be achieved as a matter of urgency because:

a. the planet's capacity to support people is being irreversibly reduced in both developing and developed countries:

thousands of millions of tonnes of soil are lost every year as a result of deforestation and poor land management;

at least 3,000 square kilometers of prime farmland disappear every year under buildings and roads in developed countries alone;

b. hundreds of millions of rural people in developing countries, including 500 million malnourished and 800 million destitute, are compelled to destroy the resources necessary to free them from starvation and poverty:

in widening swaths around their villages the rural poor strip the land of trees and shrubs for fuel so that now many communities do not have enough firewood to cook food or keep warm;

the rural poor are also obliged to burn every year 400 million tonnes of dung and crop residues badly needed to regenerate soils;

c. the energy, financial and other costs of providing goods and services are growing:

throughout the world, but especially in developing countries, siltation cuts the lifetimes of reservoirs supplying water and hydroelectricity, often by as much as half;

floods devastate settlements and crops (in India the annual cost of floods ranges from $140 million to $750 million);

d. the resource base of major industries is shrinking:

tropical forests are contracting so rapidly that by

the end of this century the remaining area of un-logged productive forest will have been halved;

the coastal support systems of many fisheries are being destroyed or polluted (in the USA the annual cost of the resulting losses is estimated at $86 million).

3. The main obstacles to achieving conservation are:

a. the belief that living resource conservation is a limited sector, rather than a process that cuts across and must be considered by all sectors;

b. the consequent failure to integrate conservation with development;

c. a development process that is often inflexible and needlessly destructive, due to inadequacies in environmental planning, a lack of rational use allocation and undue emphasis on narrow short-term interests rather than broader longer term ones;

d. the lack of a capacity to conserve, due to inadequate legislation and lack of enforcement; poor organization (notably government agencies with insufficient mandates a lack of coordination); lack of trained personnel; and a lack of basic information on priorities, on the productive and regenerative capacities of living resources, and on the trade-offs between one management option and another;

e. the lack of support for conservation, due to a lack of awareness (other than at the most superficial level) of the benefits of conservation and the responsibility to conserve among those who use or have an impact on living resources, including in many cases governments;

f. the failure to deliver conservation-based development where it is most needed, notably the rural areas of developing countries.

4. The World Conservation Strategy therefore:

a. defines living resource conservation and explains its objectives, its contribution to human survival and development and the main impediments to its achievement;

b. determines the priority requirements for achieving each of the objectives;

c. proposes national and subnational strategies to meet the priority requirements, describing a framework and principles for those strategies;

d. recommends anticipatory environmental policies, a cross-sectoral conservation policy and a broader system of national accounting in order to integrate conservation with development at the policy making level;

e. proposes an integrated method of evaluating land and water resources, supplemented by environmental assessments, as a means of improving environmental planning; and outlines a procedure for the rational allocation of land and water uses;

f. recommends reviews of legislation concerning living resources: suggests general principles for organization within government; and in particular proposes ways of improving the organizational capacities for soil conservation and for the conservation of marine living resources;

g. suggests ways of increasing the number of trained personnel; and proposes more management-oriented research and research-oriented management, so that the most urgently needed basic information is generated more quickly;

h. recommends greater public participation in planning and decision making concerning living resource use; and proposes environmental education programmes and campaigns to build support for conservation;

i. suggests ways of helping rural communities to conserve their living resources, as the essential basis of development they need.

5. In addition, the Strategy recommends international action to promote, support and (where necessary) coordinate national action, emphasizing in particular the need for:

a. stronger more comprehensive international conservation law, and increased development assistance for living resource conservation;

b. international programmes to promote the action necessary to conserve tropical forests and drylands, to protect areas essential for the preservation of genetic resources, and to conserve the global "commons"—the open ocean, the atmosphere, and Antarctica;

ċ. regional strategies to advance the conservation of shared living resources particularly with respect to international river basins and seas.

6. The *World Conservation Strategy* ends by summarizing the main requirements for sustainable development, indicating conservation priorities for the Third Development Decade.

The three initiating organizations—IUCN, UNEP, and WWF—have continued to develop and implement the goals of the strategy at the international level. In the fall of 1983 they announced a conservation plan intended to strengthen efforts to put the strategy into practice. It has seven main themes:

Mobilizing an international network for conservation action

Developing better means to monitor and analyze data

Promoting conservation as part of economic development

Conserving biological diversity

Protecting habitats and ecosystems

Maintaining a world overview of conservation concerns

Identifying key conservation issues of the future.

These themes will be the international conservation agenda of the 1980s.

In addition to prompting these international efforts, there are now strong signs that the strategy has claimed a major place in the daily thinking of national governments. More than thirty countries are already preparing national strategies modeled on the World Conservation Strategy. These efforts will help to create a sustained institutional response to the highest priorities for the protection of living resources.

But a major problem still remains. The ideas the strategy contains must penetrate to the regional and local levels, where environmental perception

has tended to be removed both from an overall strategic framework for resource conservation and from concern for development. The success of the strategy depends on strong public support. This book is intended for those who want to begin bringing the context of an international overview to their own conservation thought and action. At the local and regional levels, the strategy is most important as a framework for analyzing problems and finding solutions. It offers the opportunity to unify the work of conservation and development, whatever the scale of the immediate problem.

The discussion sections that follow each chapter are excerpted from seminars held at Tufts University, where local conservationists met with the international experts published in this book to examine the World Conservation Strategy and through it, important conservation and development issues. Contributors notes, printed at the back of this book, identify the authors of this volume and the local panelists for the discussions.

The work of preparing *Sustaining Tomorrow* was shared beyond the authors and editors. Dr. and Mrs. Harold Coolidge and Dr. James Aldrich have had important parts in carrying out this project. The production staff has been especially dedicated, even though much of the tedious detail has fallen on them; we thank Ann Gerroir, Marcie Laden, Julie Gales, and John Bolduc. Funding for this project has been provided by the World Wildlife Fund, World Wildlife Fund—US, the James G. Hanes Memorial Foundation, and the William P. Wharton Conservation Trust. Their generosity has made our work possible.

FRANCIS R. THIBODEAU
HERMANN H. FIELD

I Introduction

The World Conservation Strategy

LEE M. TALBOT

Human beings, in their quest for economic development and improvement of the quality of life, must come to terms with the reality of resource limitation and the carrying capacity of ecosystems, and must take account of the needs of future generations. This is the central message of modern conservation and of the World Conservation Strategy. Conservation is basic to human welfare and, indeed, to human survival. But it has not always been recognized as such.

Conservation has been used with many different meanings. In the sense that we are using it here, it refers to the proper use of living resources. It can be defined more precisely as the management of the biosphere to yield the greatest sustainable benefit to present generations while maintaining its potential to meet the needs and aspirations of future generations. A central part of conservation, then, is managing resources in such a way that the options for use of the same or other resources are maintained for future generations. If a species' ecosystem or ecological process is destroyed, our descendants are denied its use.

The words *conservation* and *preservation* are often used synonymously, in contrast to *utilization*. However, as I use the term *conservation*, it includes both preservation and utilization. Within this definition, *living resources* refer to components of the biosphere that reproduce themselves—flora and fauna, including microorganisms. Living resources are renewable if they are conserved, and they can be destroyed if not conserved. This fact presents an interesting paradox. Resources are generally classified in two categories: *renewable*—in other words, living resources—and *nonrenewable*—for instance, minerals. Most nonrenewable resources, such as chemicals and minerals, can be synthesized in the laboratory if they are lost in their natural state; but if renewable resources—species of living things—are exterminated, they can never be recreated. Renewable resources, then, are in the absolute sense nonrenewable, and their management must take this fact into account.

Conservation is everybody's business. But like most international problems, conservation depends on, and has its roots in, national, local, and ultimately individual actions. Consequently, the World Conservation Strategy should not be considered in abstract terms, but instead as something that must be applied locally and individually. The following discussion should be viewed as a context for setting an individual course of action to help implement the principles of the strategy, which are the principles of conservation.

Objectives of the World Conservation Strategy

Conservation has three basic objectives:

1. To maintain essential ecological processes and life support systems
2. To preserve genetic diversity
3. To ensure that the utilization of living resources, and the ecosystems in which they are found, is sustainable.

These objectives are interrelated, in the sense that each affects the others.

1. Essential ecological processes are the natural systems that are needed to maintain and sustain the living components of the biosphere in general and, more particularly, to maintain food production, health, and other necessities for human survival. These processes include the global biogeochemical cycles—such as those of nitrogen, carbon, and oxygen—and more localized phenomena, such as cy-

cling of other nutrients, soil formation, regulation of water flow, and provision of critical habitats. Because these ecological processes are vital to preserve life, they have become widely known as *life support systems*.

2. Genetic diversity refers to the genetic material in the wide range of living organisms. When a species or other class of living things is exterminated, its genetic material, and its contribution to future genetic material, is lost forever.

The question is frequently asked, Why do we need genetic diversity? Why worry about extinction of species? There is a series of compelling answers. Some species have clear direct benefit to mankind, particularly those that are harvested for food or medicines or managed for other values; insects that pollinate human food crops; or predators that prey on species considered harmful to man. Other species or varieties have a present or potential role in breeding, for example, to maintain and improve crops, livestock, timber trees, and aquatic life forms for aquaculture.

In the long run, however, perhaps the greatest value to humans of most wild species will prove to be their role in maintaining the health and stability of their ecosystems and their component ecological processes, in other words, their role in our life support system. Allowing a species to be exterminated because we do not know that it has any value to us is analogous to passengers in a hypothetical spaceship throwing out part of their life support equipment because they want more room and do not know yet what the equipment is good for.

3. Sustainability of living resources is the final objective of conservation. It is virtually a truism to say that if the utilization of a plant or animal is not sustainable, that is, if it is overharvested, the point will be reached when the species is so depleted that its value to man will be severely reduced or lost. Whales provide the classic example; overharvest has driven one species or stock after another into commercial, and in some cases biological, extinction. With the decline of stocks, the whaling industry of country after country collapsed, and the in-

FIG. 1.1. *The most significant value to humans of most
wild species will prove to be their role in their component
ecosystems. These giraffes are an integral part of a complete savanna food web that sustains many of the world's
most dramatic species. Photograph by M.J. Odell.*

dustry now survives largely on the minke whales,
the species so small that it was considered uneco-
nomical to harvest until several years ago, when
the large forms had virtually disappeared.

The growing world population requires ever-increasing amounts of protein. Significant quantities
of protein are now provided by both commercial
and subsistence harvest of wild species, particularly
marine mammals, fishes, and invertebrates, but
also birds, terrestrial mammals, and insects. Such
foods provide a major part of the animal protein intake of people in large areas of Africa, Asia, and
Latin America, and even in parts of northern Europe and America.

Human Significance
of Conservation

It is clear from the foregoing discussion that the
word *conservation*, as we use it, covers a very broad
set of environmental considerations that are of
basic human concern from two points of view.
First, conservation seeks to maintain the capability
of the earth to support life, including human life,
by maintaining the health and proper functioning
of the ecological life support system, ecological

processes, and the genetic diversity within them,
all of which are essential for human welfare and survival.

Second, for many people conservation represents an ethical imperative. This is expressed in
various ways, such as, "We have no right to destroy
any other life form"; "We have the capability to destroy other forms of life, therefore we have the responsibility to see that they are not destroyed"; "We
have not inherited the earth from our parents, we
have borrowed it from our children."

Conservation is of direct concern to all peoples
and all nations, whether or not they recognize it at
this time. Regardless of background, nationality,
type of government or political concern, and even
economic status, all are immediately involved with
human survival and welfare. And, of course, ethical concerns about conservation can also cut across
political and ideological boundaries. It follows that
conservation is important to all levels of human activity—international, national, local, and individual.

Humans' Impact on the Face of the Earth

Fundamental to conservation is the realization that human activities have a significant impact on the face of the earth, particularly on its fauna, flora, soils, and waters. It is rarely recognized, however, that these impacts extend into prehistory, probably virtually as far back as the human species itself. The human species has changed the vegetation, and consequently soil and water regimes, of the earth ever since we domesticated fire, perhaps as long as a million years ago; and our more recent agricultural practices, including shifting cultivation and grazing of domestic livestock, have further modified a large part of the world's land surface (Thomas, 1956).

My own ecological studies and other environmental work, which have taken me to over one hundred nations, have convinced me that with few exceptions the present location and composition of tropical savannas and many other grasslands is largely anthropogenic; that the same is true of many areas which are now desert and some temperate forests (as they existed before the industrial era), particularly those in western Europe and North America; and that human activities also have had much to do with the present (including preindustrial) distribution and composition of the wild fauna in many parts of the world (Talbot, 1957, 1960, 1964). For example, I believe that shifting cultivation allowed the spread of Southeast Asia's rich variety of large wild mammals (including many species of wild cattle, deer and deerlike animals, elephants, rhinos, and pigs) into areas that otherwise would have been closed tropical forests in which such animals cannot thrive or even survive.

I have long been convinced that the human activities which resulted in forest clearance, denudation of other vegetation, and desertification caused local and possibly regional or global climatic changes. For many years this view was not generally accepted. Now, at last, there is increasing evidence that anthropogenic environmental changes may have caused significant regional climatic changes, and indeed "that humans have made substantial contributions to global climate changes during the past several millennia, and perhaps over the past million years; further such changes are now under way" (Sagan, Toon, and Pollack, 1979).

From this perspective human impact on the biosphere and atmosphere is nothing new. What is new is the massive increase in the *rate* of change, caused by the exponential growth in the human population interacting with the leverage of modern technology, and the *new dimensions* to the change—such as chemical pollution—also caused by modern technology.

The Origins of Conservation

Just as human impact on the environment is not a new phenomenon, human concern about that impact is also not new. Over two millennia ago Plato eloquently recorded his anxiety about the hills of Attica in Greece, which had been denuded of their forest cover and had consequently lost their mantle soil and watercourses. They were, he wrote, "like the skeleton of a body wasted by disease" (Taylor, 1929). Further east, in the same period, protected forest areas were established in India, precursors to our national parks and reserves (Artha Shastra, c. 300 B.C.). Also in India, Emperor Asoka established the first recorded "game laws," providing protection for certain species of mammals, birds, and fish (Asoka, c. 250 B.C.). Both of these developments represented conservation actions taken in response to clear recognition of the need to control human activities to avoid harmful impact on wild living resources.

Throughout subsequent history only a few scientists and philosophers saw the changes that were taking place on the face of the earth and perceived how these changes adversely affected man; fewer still realized man's role as a causative factor. At least in Western cultures, recognition of man's voluntary causal role was clouded by the religious beliefs that such changes were expressions of "God's will." However, growing awareness of the finite limits of resources and of man's role in the environment was crystallized in the mid-1900s by writers

FIG. 1.2. *This forest in Madagascar is only one ecosystem increasingly subjected to anthropogenic changes from an expanding population. Photograph by J. J. Petter from the World Wildlife Fund.*

such as George Perkins Marsh (see, for instance, Marsh, 1864).

But even then, the intellectual insight of the few did not lead to general acceptance by the scientific establishment, much less the public, nor, consequently, did it impel meaningful action by governments. It seems to be virtually a law of nature that people are not moved to action until they see a problem clearly with their own eyes. I call this the "instant catastrophe syndrome." When change is slow it passes unnoticed. Plato referred to this phenomenon, noting that the only remaining evidence that forests had existed in Attica were the logs used in the construction of temples and that stone shrines marking long-dry springs were the only evidence of the once abundant streamflow. Only when change is so rapid that it and its consequences occur within one person's memory span may such change lead to action.

During all but the past few decades of human occupation of the earth, change has been extremely slow. The denudation of significant parts of the forest cover of Europe, China, other parts of Asia, Africa, and the Americas occurred over such long periods that many living in the now treeless regions have no concept that conditions there were ever any different. The same is true of many of the anthropogenic deserts.

However, a dramatic acceleration in the rate of change—and consequent rate of recognition of change—occurred in North America in the latter half of the nineteenth century. During the settlers' westward movement, farmland was exhausted, forests were cut and burned, and wildlife was wiped out, but there were always unlimited new lands and resources to the west. Suddenly, they came to the Pacific Ocean and realized there was no more land—they had reached the limit.

This dramatic change—the exhaustion of vast areas of prime agricultural land, clearing of once vast forestland, and virtual annihilation of formerly endless herds of buffalo and other wildlife—occurred within the memory span of those then living. Their recognition led to action, and to an unprecedented series of conservation programs, particularly for forests and wildlife. Several decades later North America experienced another instant catastrophe in the form of the dust bowl, which in turn resulted in dramatic national action to achieve soil conservation. Governmental awareness was furthered by the dust from the afflicted states, which blew over a thousand miles to Washington, D.C., where it served as a most visible reminder of the problem.

The most recent period of instant catastrophe occurred in the late 1960s and was primarily linked with pollution. Burgeoning industrialization in the absence of environmental controls led to a series of severe pollution incidents, such as the Minamata disease caused by mercury poisoning in Japan, massive tanker spills, and increasingly severe atmospheric pollution over main centers of population. These combined with modern communications, which assured that large numbers of people "saw" the situation via television, to provide initial impetus for the unprecedented global conservation actions of the past decade.

But it is important to reemphasize the value of individual actions. People became aware of the problems and took individual actions, including changing their own life-styles and purchasing habits (for instance, using biodegradable washing materials), joining or establishing citizens' environmental organizations, and seeking to assure that their governments took environment into account in their local, national, and international activities.

The Need for a Holistic Approach

The glorious photographs of earth taken from space brought some recognition that we have "only one earth" (the motto of the Stockholm conference) and that it and the resources on it are indeed finite. There has been growing scientific and public recognition of this fact, and the concurrent fact that everything is interrelated, so we need to approach the management of our environment holistically.

Yet to date most of our endeavors are fragmented, dealing with one or another problem in a largely isolated and consequently simplistic way, assuming that somehow one part of the environment is separate from, and can be dealt with apart from, the rest. Nowhere is this better illustrated than in a recent study, undertaken by the U.S. government, of the probable changes in the earth's population, resources, and environment to the year 2000. It was intended not simply to project present trends, each in isolation from the others, but to consider the synergistic interactions among different phenomena. Obviously, factors such as population, agriculture, transportation, forestry, and energy production are intimately interrelated. However, after three years of effort, the study concluded that the government did not have the capability to deal effectively with interrelationships among such different but clearly related factors in terms of their environmental implications (CEQ, 1980).

At first, this appears incredible. On further consideration, however, it is totally consistent with our cultural and scientific development. Western science operates largely on the basis of reductionism: a complex whole is divided into its simpler components, on the theory that they are individually easier to understand and that once the components are understood they can be reassembled to allow comprehension of the whole. The problem is that most effort is devoted to the first part of the process, the reduction, not to the reassembly and synthesis. The consequence is proliferation of specialties and specialists, rather than of synthesis and synthesizers.

This process is further exacerbated by the traditional academic disciplines, expressed in educational institutions divided into virtually watertight compartments, which make elegant sense from a classical point of view, but which bear little relation to the real world. The systems of academic advancement and awards, however, are based on these disciplinary divisions. One result is that if a scientist emerges from the educational system able to synthesize with scientific rigor, in other words, to deal with a holistic environment in a realistic way, it is usually in spite of the system rather than because of it. This is true not only of science, but of other academic endeavors and social aspects of life, including government, as well.

Again, the example of the U.S. government is instructive. In 1969, when Congress determined that the government should develop an institutional capability to deal with the environment, it discovered that there were around eighty individual federal agencies which either were responsible for some aspect of environmental protection or had responsibilities that affected the environment. All these agencies had some relationship with the environment and, therefore, with one another, yet there was no liaison or coordination among them and no central environmental policy. The "missions" of many of these agencies were in direct opposition to environmental protection or to those of other agencies; for example, one agency's road-building responsibilities conflicted with another's agricultural responsibilities which in turn conflicted with another's wildlife protection responsibilities. The United States ameliorated the problem with subsequent legislation establishing a national environmental policy and institutional arrangements to implement it. But this example illustrates that governments, like science and the educational system, are organized in a reductionist, compartmentalized way along the lines of individual, isolated "missions" or objectives, and hence are not well adapted to approaching environmental problems from a realistic, holistic perspective. This is a central problem facing contemporary conservation.

Necessary New Directions in Conservation

Compartmentalization has also characterized most past efforts at conservation. In large degree, conservation at both national and international levels has been reactive. A problem was perceived—usually a threat to a species or area—and conservationists responded. This approach has been responsible for the very considerable successes of conservation in the past decades, but it has little chance of lasting success against the challenges ahead.

In the first place, the reactive approach constitutes an ad hoc type of action. There is no way to assure that limited resources are applied to the highest-priority problems, rather than to the most immediately visible ones; to establish goals and focus a wide array of efforts; or to establish benchmarks by which achievement can be judged.

Second, the reactive approach virtually always focuses on the *effect* rather than the *cause*, the symptom rather than the sickness. If a forest bird is threatened by forest clearance, the reactive approach would be to pass laws to protect the bird or establish a small reserve for it. This amounts to putting a bandage on the symptom of a chronic illness. Unless something is done about the base causes of the clearance, eventually the forest will be gone, with all its other organisms, leaving the bird's reserve an island of trees, likely soon to be lost through ecological change or economic pressure.

A further major weakness in the reactive approach is its focus on *cure* rather than *prevention*. In the contemporary world, once a conservation problem is occurring, it is extremely difficult to do anything meaningful about it. Government planning and development activities illustrate the point. There may be years between the initial proposal for a given development activity or other agency action and its initiation. By the time work actually starts, engineering and economic studies have been made; budgets developed and approved; national and, as necessary, international agreements concluded; contracts let; and people hired or assigned. A major investment in time and money has already been made, and the bureaucratic process has swung into motion. Stopping or significantly altering the action at that point is extremely costly and difficult, if possible at all. Clearly, the time to have acted would have been at the start. Had conservation considerations been brought in at a much earlier stage in the planning or decision making, the action could have been redirected both to benefit the development objective and to avoid the conservation problem rather than reacting to it it after the fact.

This, in turn, leads to what is probably the most serious problem of the reactive approach—that it is

FIG. 1.3. *Morelet's crocodiles, and thousands of other species, will continue to be in jeopardy as long as the basic problems that endanger them remain. Photograph by J. H. Powell, Jr. from the World Wildlife Fund.*

virtually always perceived as antidevelopment, against human welfare. Consequently, it places the conservationists outside of, or in opposition to, the mainstream of human activity, denying them the political, economic, and moral support necessary to achieve lasting goals.

Recognizing these weaknesses of past approaches to conservation, while appreciating what they have accomplished, the International Union for Conservation of Nature and Natural Resources (IUCN) is leading a growing international movement to redirect conservation endeavors. Its new directions may be described as programmatic, goal oriented with a strategic approach to achieving high-priority objectives, focused on causes as well as effects, and concentrated on preventing problems before they occur. Above all, the new emphasis is on the vital importance of conservation to human welfare.

Necessary New Directions in Development

While the problems described above have posed major obstacles to achievement of *conservation goals*, a parallel set of problems have equally ob-

structed achievement of *development goals*. *Development* in this sense refers to the broad array of activities—local, national, and international—intended to satisfy human needs and improve the quality of human life. To be successful, development must not only prosper in the short run but also be sustainable, economically and ecologically.

Unfortunately, much development worldwide has not been successful in these terms. Some activities are themselves short-lived because of inherent ecological errors. One example is range management development that does not include control of livestock numbers—it leads to overgrazing and collapse of the resource. Other projects deal with part of a system and are defeated by problems elsewhere in the system. Examples include hydroelectric or flood control dams whose essential watershed areas were not protected and so became denuded, leading to erosion, which filled the dams with mud instead of water. Projects that appear successful in themselves may adversely affect the sustainability

FIG. 1.4. *These Egyptian solar collectors exemplify economically and ecologically sustainable development projects. Photograph by M. J. Odell.*

of the environment as a whole for people. Construction of industries, transportation systems, or housing on prime farmland; unsuitable agricultural development leading to the loss of cropland; and commercial overexploitation of fisheries and forests are examples. In any event, there has been so much ecological backlash from development projects which did not take conservation requirements into account, that in all too many cases such projects have reduced, rather than increased, the human carrying capacity of their areas, consequently decreasing, rather than enhancing, human welfare.

A further problem has involved the type and direction of development assistance. For many years much international development was based on the "trickle down" theory, which assumed that development which assisted industry and the upper economic classes in a developing country would trickle down to the poorest citizens. The fallacy of this approach is now being recognized, and international development is at last being directed to "the poorest of the poor."

Whatever the individual causes and despite the millions of dollars spent on development in the past decades, some 500 million people—one-eighth of the world's population—are malnour-

ished (FAO, 1977) and over one-quarter of the population is regarded as destitute or poor (World Bank, 1982). Far more serious are the indications that the life support system of the earth—its carrying capacity for humans—is being seriously eroded.

What Lies Ahead?

Recently there has been a series of attempts to analyze the present conservation situation on a regional or global basis, and to project current trends into the near future. These efforts have included individual scholars' models, institutional or organizational projects, governmental programs, and international studies and conferences. Most studies agree on the present conditions, but, as would be expected, there are differences in their future projections. The major variations appear to be associated with how comprehensive the analyses were, how much they took into account interrelationships among major factors (such as energy costs' influence on agricultural practices or the impact of forest reduction on water regimes, and consequently on agriculture and food productivity), and, above all, how much they considered the environmental impacts of projected increases in human populations and their activities.

However, the most recent and environmentally comprehensive studies lead to the following conclusions about some conditions in the year 2000—if present policies and activities proceed relatively unchanged:

Tropical lowland forests will be largely gone in most areas, with some few exceptions, such as parts of the Amazon basin. Even the most optimistic projections indicate that close to half of the present tropical forests will be gone.

Other forests (high-altitude forests, open forests, and woodlands) in the tropics and subtropics will be greatly reduced, and gone in many areas.

All vegetation over vast areas will be severely denuded. The world's drylands are being degraded at a rate of almost 60,000 square kilometers a year; the United Nations regards 20 million square kilometers of land as being on the brink of desertification.

Roughly one-third of the world's present cropland will be gone—lost to erosion, bad irrigation, encroachment of desert, and replacement by cities, transportation systems, and industry.

Loss of the forests and other vegetation will destroy the watershed and interrupt the water regimes, bring floods in the wet seasons and droughts in the dry. These changes, in turn, will reduce agricultural productivity of much of the remaining cropland.

Loss of habitat, particularly tropical forests, plus overexploitation will result in the extinction of between 15% and 20% of all living species of plants and animals. At present, over 1,000 vertebrate species and an estimated 25,000 species of plants are known to be threatened with extinction.

Because of overfishing and near-shore habitat damage, fishery yields will continue to decline, and many of the present major fisheries may collapse.

The human population will increase by at least 50%, from around 4 billion people today to over 6 billion by the year 2000. Simply to feed these people at present levels will require an increase of 50% in current production from agriculture, fisheries, and wildlife. Yet the factors cited above will reduce productivity in much of the world. Therefore, there is a strong likelihood of substantial actual reduction of available food, on a per capita basis, in some parts of the world, particularly southern Asia and Africa. This sobering warning was also central to the message of the Brandt Commission Report (ICIDI, 1980).

Loss of forests, increased industrialization, and increasing desertification will affect the climate on a global basis. There is some dispute about what will eventually happen, but it seems likely that in the near future the climatic fluctuations or instability of the 1970s (relative to the stable climate of the previous decades) will continue and perhaps worsen. This will have further unfavorable effects on food production, particularly in the developing world. The longer-term impacts, particularly a global warming caused by increased carbon dioxide, could be nothing short of catastrophic (CEQ, 1980).

This scenario is a grim one, and if it is reasonably close to the truth, it points up several factors of great significance for conservation and development. First, it would result in mass human starvation, and the world's political, social, and economic systems would not allow this to happen, at least not without very severe disruptions and instability.

Second, the scenario assumes no change in present policies and actions. Consequently, it constitutes the most powerful argument for the critical need to make changes to avoid these projected results.

Third, the principal problems are ecological—degradation of the life support system. These are intimately linked with development in two ways: One is ecologically unsound development, which is not sustainable and which in the long run lowers the carrying capacity for humans and defeats its own original purpose. The other way is lack of development. If present patterns of development continue, increasing numbers of people will be living at the bare subsistence level and will have no choice but to denude the land for fuel and grazing and overexploit wildlife and whatever other resources are available to them simply to survive.

The Essential Link between Conservation and Development

Clearly, conservation and development are essential for each other. Unless development conforms to conservation principles, it is not sustainable and human welfare is not served. At the same time, unless there is adequate development—which must be ecologically sound—conservation will be undermined by the subsistence requirements of the increasing populations needing development assistance. The essential linkage between development and conservation is a new concept for most developers and conservationists, and, indeed, that is one reason the global situation is as unsatisfactory as it is today.

Of the many factors involved, the main obstacles to achieving conservation and ecologically sustainable development are probably the following:

1. The belief of conservationists and others that conservation is a separate, isolated concern, rather than one that must be integrated throughout human endeavors
2. The failure to integrate conservation into development at all stages and all levels of development activity—local, national, and international
3. A development process that has been narrow in terms of immediate goals, generally inflexible (in other words, bureaucratic), and ecologically damaging because of failure to incorporate environmental considerations
4. A development process that essentially did not aid the increasing numbers of rural poor
5. Inadequate conservation capabilities of governments, in terms of factors such as insufficient mandate and legislation, inadequately trained personnel, and insufficient information on conservation processes and needs
6. Lack of awareness of the need for conservation

and consequent lack of support for conservation among the public, industry, governments, and international institutions.

Development of the World Conservation Strategy

To meet these multiple challenges, a new approach was needed. In recognition of this need, IUCN, the U.N. Environment Program (UNEP), and World Wildlife Fund (WWF) collaborated in the development of the World Conservation Strategy (IUCN, 1980). The IUCN prepared the document, and UNEP and WWF provided financial support and contributed to the evolution of its basic themes and structure. The breadth of the participation by these three organizations is most significant. We believe that this is the first time a project of this magnitude has been undertaken in the field of conservation.

The World Conservation Strategy presents a clear statement of conservation priorities and a broad plan for achieving them. It is a strategy in the military sense, in that it defines goals, assigns priorities, and lays out a framework for specified action to accomplish the goals. It also identifies major obstacles. And it outlines specific steps—policy decisions and other actions—to accomplish the goals at worldwide, regional, and national levels.

The strategy is aimed at three main groups of users:

1. Government policy makers and their advisers. For them, the document recommends ways to overcome the main obstacles to conservation and provides specific guidance on what action is most important. The strategy is designed to be relevant to any level of government with significant responsibilities for planning and managing the use of living resources.

2. Development practitioners, including aid agencies, industry and commerce, and trade unions. The strategy demonstrates the need for conservation to improve the prospects of sustainable development and identifies ways of integrating conservation into the development process.

3. Conservationists and others in the private sector directly concerned with living resources. The strategy indicates to them the areas where action is most urgently needed and where it can be expected to yield the greatest and most lasting benefits. It also proposes ways conservation can participate in the development process, which is central to the success of the entire venture.

The development of the strategy itself provides some indication of one way the final document should work. Discussions about the need for a strategic approach were initiated within IUCN in 1969. Plans for its actual development were started in 1975, and work on the first draft began after that. In all, there were four "official" drafts, the final version, and several intermediate efforts. Each draft was submitted to the full IUCN membership and also to nearly a thousand scientists and other advisers for comment. Several international meetings were held to review and comment on the then current drafts, and a formal IUCN advisory committee met frequently. The final draft was also approved and formally endorsed by the sponsors, the U.N. Educational, Scientific, and Cultural Organization, and the U.N. Food and Agriculture Organization.

The first draft was essentially a wildlife conservation textbook. At that time many conservationists regarded development as the enemy to be opposed, and many developers, for their part, regarded conservationists as at best something to be ignored, at worst an obstacle to progress. Each draft brought the two sides closer. Each was a process of education. The final draft represents a consensus between the practitioners of conservation and development, a consensus that would not have been possible without the educational experience which development of the various drafts provided. In a sense, the very existence of the present strategy with the formal support it has is proof of the document's basic premise concerning the essential interdependence of conservation and development, and evidence that the "two sides" can cooperate—as indeed they must.

The strategy has appeared when there is growing international recognition of the interdependence of conservation and development. For example,

the month before the strategy's release the Brandt Commission completed its two years' work and released its report, called *North-South: A Programme for Survival*. The report analyzes the world's economic and social predicament as it affects the Third World and concludes with a set of far-reaching proposals for the reform and restructuring of the world system, which in the commission's view are essential to avert disaster and are in the mutual interests of both north and south. The report states, "Few threats to peace and the survival of the human community are greater than those posed by the prospects of cumulative and irreversible degradation of the biosphere on which human life depends. . . . It can no longer be argued that protection of the environment is an obstacle to development. On the contrary, the care of the natural environment is an essential aspect of development" (ICIDI, 1980).

Also that month the heads of the World Bank and eight regional development banks and development assistance agencies met in New York and signed the Declaration of Environmental Policies and Procedures relating to Economic Development.[1] This is a very powerful document aimed at assuring that the development activities of the signatories adequately take environment into account, because they are "convinced that in the long run environmental protection and economic development are not only compatible but interdependent and mutually reinforcing." A further consequence of this declaration was a subsequent meeting in Berlin of national development agencies to consider similar action to incorporate environmental factors in bilateral assistance programs.

The U.N. General Assembly, summed up this growing international awareness when it noted that "the [World Conservation] Strategy is based on a clear conception of conservation as a major factor

---

[1] This declaration was signed at the United Nations in New York on 1 February 1980 by the International Bank for Reconstruction and Development, the U.N. Development Programme, the African Development Bank, the Arab Bank for Development in Africa, the Asian Development Bank, the Caribbean Development Bank, the Commission of the European Communities/European Development Fund, the Inter-American Development Bank, and the Organization of American States.

in sustaining the much needed development especially in developing countries" and unanimously adopted a decision that "welcomes the collaboration between UNEP, IUCN and WWF in the development of guidelines to help governments in the management of their living resources through the formulation of a *World Conservation Strategy* to be launched in March 1980" (U.N. General Assembly, 1979).

Normally a program of this magnitude would be initiated via a U.N. conference, such as the conferences on food, population, and habitat. However, it was considered more appropriate to introduce the World Conservation Strategy in individual countries throughout the world in a coordinated "launch," held on 5 March 1980. Because the launch was a simultaneous press conference in thirty-five nations' capitals, it constituted an unprecedented media event, which itself served to bring it to the attention of many people throughout the world. Further, because the strategy was the result of unique cooperation between the nongovernmental and governmental communities, it was not purely a U.N. operation and, therefore, was less appropriate for the U.N. conference format. Finally, conservation is basically nonpolitical, it is a bridge across the political and ideological differences that often separate people, and the strategy's launch was intended to emphasize that. Consequently, events were held in the capitals of nations that covered the entire spectrum of political systems, levels of industrial development, and geographic locations—including Peking, Moscow, New Delhi, Brasilia, Amman, Caracas, Nairobi, London, Washington, Jakarta, Bangkok, and Canberra.

Most of the launch events were presided over by heads of state and government, who emphasized the level of strategy endorsement by the nations involved. The U.N. secretary general, Dr. Kurt Waldheim, described the strategy as a "remarkable pooling of international resources which has resulted in an unprecedented degree of agreement on what should be done to ensure the proper management and optimal use of the world's living resources, not only for ourselves but also for future generations" (1980). Many governments and intergovernmental organizations including Australia, Egypt, India, Indonesia, Jordan, Kenya, the Peo-

ple's Republic of China, Spain, the United States, the European Parliament, the European Economic Community, and the Organization of American States publicized decisions and programs in direct support of the strategy.

For example, several governments, among them India, the USSR, New Zealand, and Thailand, announced development of national conservation strategies, one of the key recommendations of the world strategy. The European Parliament created a European Environmental Fund. The People's Republic of China declared the month the strategy was launched National Conservation Month and carried out intensive educational programs that reached all levels of their society—bringing the strategy to one-quarter of the world's population in a single action. Since its launch the strategy has been endorsed by an impressive array of additional governments and national and international institutions. Formal actions to implement it are under way in over forty nations, and it has been translated into many languages. The strategy's sponsors have undertaken a joint follow-up project to keep the strategy on the world's agenda and see its fuller implementation.

However, much of the response to the strategy so far has been at high governmental levels. This is essential, but there is also a very great need for action at the individual and local levels. In this sense the World Conservation Strategy carries as much of a message to the individual citizen as it does to world leaders, and a major challenge now is to translate its guidance into local action.

The strategy exists, it has been introduced to the world, and it already has achieved significant results. In one sense, then, a major effort has been culminated. In a broader sense, however, we are at the start of a new phase in conservation. For although the strategy is the most ambitious effort ever undertaken in international conservation, from a historical perspective it is simply a part of the ongoing process. The challenge now is to make the strategy work, to see that its recommendations are implemented, and, most important, to see that it does serve as a focus for cooperation among all segments of world society to achieve common goals that will maintain a world in which human welfare—and survival—is possible.

An Introduction to World Conservation

RAYMOND F. DASMANN

My comments will concern my own experience with the International Union for Conservation of Nature and Natural Resources (IUCN), the background of the World Conservation Strategy, and the new concepts of development being brought forward today. I worked in Africa from 1959 to 1961, trying to institute game ranching as an economic activity in what was then the Federation of Rhodesia and Nyasaland (Dasmann, 1964). That experience was the first time I was involved with a world conservation problem. Not only was wildlife being wiped out at a rapid rate over much of Central Africa, but the entire political situation was becoming unstable. Nothing focuses one's attention on social problems more than a riot—and there were riots at that time. It has taken a long time for the social problems in that part of the world to be even partially solved. It will take a longer time to bring conservation issues under control. But I developed a sense of urgency when I saw these problems firsthand, and I have carried this urgency through my subsequent experiences.

IUCN and International Environmental Conservation

My actual working association with IUCN began in 1966, when I joined the Conservation Foundation in Washington, D.C. The foundation was almost unique. In 1966 most people had not even heard the word *ecology*, but the Conservation Foundation had five professional ecologists on its staff, more than any university at that time. Frank Fraser Darling, one of the original ecologists; William Vogt; and Edward Graham were outstanding. The foundation was associated with virtually every professional ecologist in the country. Through Harold Coolidge and Russell Train, it worked particularly closely with IUCN. One of my first duties with the Conservation Foundation involved working with the U.N. Educational, Scientific, and Cultural Organization (UNESCO) on the preparation of a draft document for the Biosphere Conference, to be held in 1968 (Dasmann, 1968). This meeting led to the preparation of what is now the Man and the Biosphere Program, or MAB. The Biosphere Conference brought together experts from virtually every country in the world and for the first time focused international attention on problems of the environment (UNESCO, 1968). When the MAB program was finally approved by the UNESCO General Conference in 1971, it was on a considerably smaller scale than some of us had hoped. Nevertheless, it was the first intergovernmental program directed at environmental concerns (UNESCO, 1971).

The IUCN still works with other international organizations to initiate such activities. Often its work is behind the scenes. When a program is launched and all the credits are written, IUCN usually cannot be mentioned because of its unofficial role. The union played a major role in getting conservation into the MAB program. Without IUCN, I think MAB would have been entirely directed toward research, but because of its lobbying and other activities, MAB has developed an important conservation aspect. The establishment of biosphere reserves around the world and the emphasis within MAB on the conservation of genetic resources are to the credit of IUCN.

A second job with which I became involved while working for the Conservation Foundation was preparation of the background document, to be issued by the secretary general of the United Nations, calling for the first U.N. Conference on the Human Environment. That conference was held in Stockholm in 1972 and resulted, among other things, in the formation of the U.N. Environment Programme (UNEP). I went into that conference with hopes that we could really move ahead, but the results were disillusioning. It seemed to bring out the worst in the delegates—a tremendous amount of wrangling among Third World, First World, and Second World, with delegations storm-

ing out in a rage and coming back much later, only slowly being led into any kind of agreement. And yet, although it did not save the world and did not change things immediately, the Stockholm conference did have some positive results, not only in forming the UNEP, but in forcing what was practically the first serious look at environmental problems by many Third World governments (U.N. General Assembly, 1972).

The IUCN was again working behind the scenes. It was not among the official participants; it was an NGO, a nongovernmental organization. Yet the whole environment program would have taken a much different turn if representatives of IUCN had not been constantly working to get nature conservation and an interest in the natural environment built into UNEP. Otherwise it might have directed its interests toward pollution, land use, human habitats, and other more urban concerns.

My direct involvement with IUCN started in 1970 when I went to Morges, Switzerland, as senior ecologist on the staff. At that time IUCN had a new organization, new leadership, and a large grant from the Ford Foundation, which may have been the most important of the three changes. The Ford Foundation had emphasized the need for more centralization, more control out of Morges over what was going on. Central control was instituted, but it was to have all kinds of unfortunate repercussions years later, and it brought great dissatisfaction among the membership.

When we started work in 1970, we wanted to look first at what had been done up to that point, particularly at what projects the World Wildlife Fund (WWF) had financed, what effects those projects had, which worked, which did not, what expenditures were likely to bring results, and where money should go in the future. We discovered reluctance at WWF—a reluctance shared by virtually all international organizations—to review past successes and failures.

We also noticed an obvious imbalance in the ways funds were spent. Certain countries, Kenya or Tanzania for example, received substantial support; others, such as those in West Africa, received virtually none. Indonesia could always get grants from the Dutch WWF; Burma did not get grants from anyone. India was considered extremely important;

FIG. 2.1. *Whooping cranes in flight. U.S. and Canadian biologists have been successful in their publicly supported efforts to rebuild the population of this species, which was at the edge of extinction. Concern for a single species often leads to a broader environmental awareness. Photograph from Massachusetts Audubon Society.*

Southwest Asia received virtually nothing. This is not entirely bad, however. You have to use the opportunities you have. If you find a nation that is willing to work at conservation, it is far better to give it the support it needs. Later you can try to encourage reluctant nations. We could have put a lot of money into the Philippines in the 1960s and nothing would have been accomplished, whereas money going into Kenya produced good results. Political factors must be considered in planning a conservation strategy.

Another thing we noticed was that certain species—the large, spectacular ones, or the soft, furry ones, or the brilliantly colored ones—tended to get support; most other species did not. At first I thought that was an imbalance that needed correcting, but since then I have changed my thinking. Particularly, I have learned from the whale conservation effort that if you can get a symbol, something people are really interested in and are willing to put effort behind—a fund-raising animal, an Animal of the Year—you can accomplish a lot of conservation for other species through that one emblem.

But looking back on the 1960s, we did find a need to review and reorganize IUCN activities. For example, IUCN had put out a world directory of national parks, which was a good beginning (Harroy, 1971). Yet it only included certain categories of protected areas. We wanted to look at all protected areas and devise a system of classification for them so we could see which ones were actually accomplishing conservation, regardless of what they were called (Dasmann, 1973a). We also decided to develop a classification system for the various wild areas of the world on the basis of the species they contained. The United States and all of North America had long been looked at from this point of view, and a biotic province system had been mapped out and described for the continent. But this had not been done on a worldwide scale. Such a system, which would indicate the natural areas of the world—the major divisions according to species and natural communities—would be useful for evaluating conservation needs and for looking at conservation effectiveness. So we put a considerable amount of energy into defining and mapping biotic provinces (Dasmann, 1973b; Udvardy, 1975). The final activity, one that took much time and attention, was the updating and revision of the Red Data Books on endangered species to increase their scope so that we would know not only what species were endangered, threatened, or rare but much more about their ecology, which is what we

FIG. 2.2. *If current rates of land degradation continue, close to one-third of the world's arable land (symbolized by the stalk of grain) will be destroyed in the next twenty years. In the same period (at present rates of clearance) the remaining area of unlogged productive tropical forest will be halved and the world population will increase by almost 50%. The imbalance will be even greater by the year 2020. Illustration from IUCN, 1980.*

actually protect in parks and reserves (IUCN, 1972). All of this information was to feed into the World Conservation Strategy.

Inception of the World Conservation Strategy

We were eventually able to establish an intercommissional marine conservation program within IUCN, which brought together some of the world's best marine scientists. This program, under the leadership of G. Carlton Ray of Johns Hopkins University, gave continuing emphasis to the needs for planning and programming, for strategy and tactics. I do not think there was a single meeting of IUCN during the 1970s where planning, programming, strategy, and tactics were not brought to the attention of all the commissions of IUCN by the marine steering committee. I believe this was one impetus that led to the strategic approach to world conservation problems. It is logical that this should come from marine specialists. They deal with 70% of the world's surface. They encounter conservation problems that are often an order of magnitude greater than those of terrestrial conservation. Marine ecosystems are so interrelated, you really cannot do one thing in one place without affecting, and being affected by, everything. You must develop a strategy; you must have priorities. These realizations were an important part of the background for the World Conservation Strategy (IUCN, 1976).

But more significant than any of these activities

was the shift in emphasis at IUCN toward a concern for economic development. Conservation organizations did not usually take much interest in economic development at that time. It was considered "the enemy" in a sense. If you were for conservation, you were not for development. In many conservation circles this was taken for granted. But this attitude was to change. To go back once more to the Conservation Foundation, in 1967 the foundation began working with Barry Commoner's group in St. Louis on a conference that would bring together experts from all over the world to look at the effects of international development programs on the world environment. This conference, in 1968, reviewed the costs and benefits of a number of development schemes supported by international agencies, including river development—the big Aswan Dam in Egypt for example—agricultural development, irrigation schemes, the green revolution, and the development and use of pesticides. All of these activities had been supported by international agencies.

The transactions were eventually published in a book called *The Careless Technology*, edited by M. Taghi Farvar of Iran and John Milton of the Conservation Foundation (1972). But the real results of the conference were measured by the increased attention development agencies began to give to environmental problems. The agencies were invited to the conference, and they took an active part in it. The IUCN was again very much involved in both the planning and execution of this conference. The conference did not seek areas in which things had gone wrong, in fact a good deal of our effort went toward finding examples of international development projects that did not produce adverse environmental consequences. But we ended up examining a series of international environmental misadventures, which had cost hundreds of millions of dollars.

After this conference IUCN and the Conservation Foundation decided to pursue the matter further, in an attempt to provide development agencies with the kind of ecological knowledge they would need to avoid such mistakes in the future. A meeting was held in Rome in 1970 to bring development agency people together to discover what they needed and wanted us to do. This resulted in the preparation of a book, a series of papers, and workshops, all having to do with guidelines or ecological principles for economic development (Dasmann, Milton, and Freeman, 1973). The IUCN was almost the only conservation organization talking about the integration of conservation and development at that time. However, after the Stockholm conference, UNEP was essentially told by the developing countries that they were not interested in saving the environment unless this effort was tied to economic development. UNEP found that IUCN was in a strong position to work with it toward this goal. Thereafter, the link between conservation and development was even more strongly emphasized within IUCN. IUCN began to look at new concepts of development, and specifically the concept of ecodevelopment, which had its origin with UNEP but was related to many of the themes IUCN had been working on.

Ecodevelopment

What is *ecodevelopment*? I think the term has probably been defined best in a background paper prepared for the South Pacific Regional Environment Program (South Pacific Commission and South Pacific Bureau for Economic Cooperation, 1977). It points out that ecodevelopment has three principal components.

The first is that development must be directed toward meeting the *basic needs*—food, clothing, and shelter—of the poorest people everywhere before it pays attention to the wants or luxuries of the well to do. Now this seems reasonable, but it happens to be a complete reverse of the way most development has taken place. Most development money has been brought in at the top—at high government levels—in the belief that if those at the top benefit, something will trickle down to those at the bottom. But not much ever trickles. It goes down a little way and then dries up. An approach directed toward meeting basic needs means putting development money in at the bottom, at the grass roots level, for rural development or local development. It means getting the expertise, the equipment, whatever is needed to the people who need it and will use it and not giving it to the people who may be in a position to accept it most easily.

The second component of ecodevelopment is *self-reliance*. Development must encourage local self-reliance, not increase dependency on a former colonial power, metropolitan country, industrial center, or any distant agency. It must increase local self-reliance, including relative self-sufficiency in some essentials such as food, but not total self-sufficiency in a sense that rules out the benefits of trade. Such self-reliance should be built from local knowledge, traditions, and skills, rather than the transfer of unsuited technology. It is a self-reliance based on appropriate technology—appropriate to the environment and appropriate to the background and knowledge of the people.

Finally, development must enhance *ecological sustainability*; it must be in a symbiotic relationship with nature. Development must operate within the constraints of local ecosystems. It must maintain and encourage biotic diversity as an insurance for the future. Essentially, it must be ecologically sustainable in terms of local ecosystems and the biosphere. This is where the conservation element enters ecodevelopment.

This form of development has been picked up under different names. Sometimes it is called *rural*

FIG. 2.3. *Appropriate education, such as this agricultural program in Nepal, provides people with skills that help them maintain their self-sufficiency. Photograph by M.J. Odell.*

development. UNESCO likes to call it *endogenous development*. Others call it *local development*. The publications of the International Foundation for Development Alternatives give a new definition of development:

Some still consider that development refers to things and can be reduced to capital accumulation, economic growth and economic restructuring. But development fundamentally refers to human beings, the whole man, the whole woman. It is a human experience synonymous with the fulfillment of individual mental, emotional and physical potentiality. The society, its economy and polity ought to be organized in such a manner as to maximize for the individual the opportunities for self-fulfillment. There is development when people and their communities act as subjects and are not acted upon as objects, assert their autonomy, self-reliance and self-confidence, when they set out and carry out projects. To develop is to be or to become, not to have (IFDA, 1979).

This is the ecodevelopment approach in different terms.

In the September 1980 issue of *Scientific American* devoted entirely to economic development, the lead article by K.K.S. Dadzie of India has another wording of the same notion:

Development is the unfolding of people's individual and social imaginations in defining goals and inventing ways to approach them. Development is the continuing process of the liberation of peoples in society. There is development when they are able to assert their autonomy, and in self-reliance carry out activities of interest to them (1980).

It is in this context that we need to think of the *sustainable development* mentioned in the World Conservation Strategy. Not the old big, heavy, capital-intensive development, but development that gets at where people live, that helps them to live in balance with nature, that helps them once more to understand and appreciate what they always have understood and appreciated in the past, that helps them recover the things they have lost under the impact of sudden growth and change. The World Conservation Strategy is obviously not complete. It is a remarkable document, but it is only a beginning. We have to think of the strategy as a dynamic, changing process.

Particularly, there must be regional and local strategies for the World Conservation Strategy's implementation. A strategy is needed for each biotic province, each province subdivision, all the bioregions of the world. Each area needs a development strategy and a conservation strategy linked to each other. A bioregional strategy may be more important than national, state, or county strategies, because political boundaries change but bioregional boundaries either do not move or only change very slowly. Whatever government is in charge will have the same responsibility for the bioregion, whether it be in ten years or in ten thousand.

In developed countries we need *redevelopment* strategies. We need to put these countries on their feet in such a way that we can say, "Now it's going to last, now things are going to work." We have to redevelop our own countries along sound ecological lines. This is our biggest challenge. It is more important that we do this than that we give a little technical assistance or a bit of money here and there to other countries, although that is important too. In the long run I think if we can learn to live on our own resources, learn to develop our own self-reliance and not have to exploit other parts of the world, we will contribute most toward the future of the planet. I hope that this will happen as the World Conservation Strategy develops over the years.

Discussion

QUESTION. I would like to go beyond the problem of single issues, people becoming excited about whales for example. Many people wish that the Audubon Society would get back to "birds" more. At the society's first meeting this year, there was a woman wearing alligator shoes. It occurred to me that she was concerned with birds but that she did not understand the global context of the environmental movement. How does one get beyond the single issue to teach people that this is a worldwide thing? How does one educate people to a larger concept of ecology?

ANSWER. That is a tough question, and one that I deal with on a daily basis. There are no simple answers, if there were we would be solving all the environmental problems we have. I think you have to deal with people where they are. You cannot get everyone to stop wearing alligator shoes in one fell swoop. You have to keep hammering away at an issue like that and relating it to the broader issue of species protection. The whole commerce in products made from endangered animals is an enormous problem. If people were not buying ivory necklaces, or rhinoceros-horn dagger handles, or whatever, there would be no market for these items.

My understanding of how people get involved with the environment is that their first attraction is at an aesthetic level. This is one of the reasons the Audubon Society is lucky; it has a way to capture interest. If people become fascinated by some aspect of the natural world, like wild flowers or colorful birds, we can tell them more. The next step is that people go back to a place where they saw a particular bird a year ago and, behold, there is a housing development. Then they begin to realize that there is a larger problem. That is when we can say, "Yes, it is a problem. This is what needs to be done." There will always be people who are not willing to take the next step. They can find somewhere else to see herons if the first six places they used to go are destroyed. But some people do go further, first aesthetically, and then intellectually.

QUESTION. On a global basis how do we find a scientific grounding for ecodevelopment policies? Do we know that the policies we are formulating are based on the best science available and the best science that will ever be available?

ANSWER. We can never know that. The worst of it is that there are so many problems waiting for the best possible answers. If we hold out for those answers, it will be too late. The environment is deteriorating so rapidly that we should launch the long-term research needed to get the answers to the questions, but we cannot wait the ten or twenty years it will take to get answers from that research. By then the questions will have answered themselves. The catastrophe will either have occurred or not. It is that close. We must operate on the belief that the worst is most likely to happen if we continue on our present course. That does not mean we will save all the mangroves everywhere on earth, because there are a lot of mangroves in some countries. But it does mean that we cannot destroy the last 50% of them. In countries such as the United States, where development has already gone so far, we are almost at the point of saying, "No more." The little we have left is probably less than we need.

QUESTION. In your definition of ecodevelopment you stated that a large part of increased self-reliance is self-sufficiency in food production. But many nations simply do not have the natural resources for major increases in production, whether they be available land or available water. In these cases how can self-sufficiency develop? Isn't that kind of independence a political goal, in many cases, rather than an economic or ecological one?

ANSWER. I am not sure I agree with you. Self-sufficiency in food production is not possible in some nations using present large-scale agricultural techniques, but it is quite possible in almost any place using a different approach. The definition of ecodevelopment is, of course, political and economic. It is not just an ecological concept. Ecodevelopment is development for people, rather than development for the benefit of international agencies or multinational corporations. That makes it an approach that is fundamentally political. The amazing thing is that something as inherently political as ecodevelopment gets the stamp and blessing of virtually every country in the world.

QUESTION. How much of the international conservation effort is based in the USSR and China?

ANSWER. The Soviets have been very active in IUCN for many years. They have worked hard for the establishment of reserves, national parks and their equivalents, and protected areas. They are very active in species conservation, particularly through economic use of those species.

However, when you examine the environmental problems in the USSR and the United States, you find the same culprit. The same actors, wearing different hats, are doing the same things, whether polluting Lake Baikal or polluting Lake Tahoe. Some do it in

the name of profit and some do it in the name of national power and prestige, but you find the same kinds of problems in both countries. There are far more similarities than differences.

China joined IUCN quite recently as a state member. But during the years it was "undercover" it had not given up the idea of conservation. Many of its old reserves were protected, and a large number of them are still there. I expect to see the Chinese playing a much more active role at the international level in the future.

QUESTION. I think there has been much more recognition than there used to be of the significance of NGOs (nongovernmental organizations). I know that the U.N. Environment Program pays a good deal more attention to what the NGOs want to do when it is setting its agenda, because they have more long-range consistency than governments. There is a healthy upward surge in NGOs' influence, but, of course, they do not have anywhere near the resources available to governments. Do you see much hope for increasing the importance of the NGOs?

ANSWER. Private money supporting a private group can work in ways that governments cannot. Any international governmental agency, UNESCO for example, has to work through the governments of the countries involved to filter resources down to the people who really need them. The chances of reaching those people are pretty slim. In contrast, an organization like IUCN can send in an action team that does not have to go through governments at all. It can work directly with local nongovernmental groups, go straight to the people—if it is planned right. My students from the University of California work directly with villages that need help. The Peace Corps does this as well; although it is a governmental organization it behaves as an NGO to a large degree. There are many of these groups. I feel a strong need to strengthen the private sectors of both the conservation movement and the development movement so that we are not as subject to the results of elections. We can bypass the top level of government. Many times we can ignore the bureaucracy and corruption entirely and get down to where the aid is really needed.

QUESTION. I am bothered by what I hear as "loaded terms" and may not be considered a conservationist by some because I am always worried about the lines drawn between development and conservation. In task forces I have sat on, there have always been very dangerous divisions between conservationists and developers. Each group has strange ideas about what the other is "for." I come from a compromising world, so I find myself interested in what both the conservationists and the developers are saying. Do you have architects, planners, or engineers in IUCN, people more given to that side of the world? How do you go about developing a symbiotic relationship with them when the tendency seems to be for these two camps to war with each other? It really has become a white hat–black hat situation, either you are identified with an animal or you are identified with profit. Somehow it seems to me that human beings get left out of these battles.

ANSWER. Your question will find its answer in a document that is not yet written—a world development strategy. Such a strategy would have to be inclined toward conservation, just as conservation is inclined toward development. The link between the two has to work. There is no other way. People are not going to get involved in environmental questions if they are starving. They are not going to get involved if the circumstances of their lives are so overwhelming that they cannot worry about long-term problems. Conservationists are beginning to realize this, but even the World Conservation Strategy does not give this issue enough attention. It only speculates a bit about development in an ideal world, a world where all governments are benign and interested in the welfare of all their people, a world where the multinational corporations only need some better information. A more realistic approach to development is going to emerge in the coming years. The document based on it will go much further than the present strategy.

QUESTION. I keep thinking of René Dubos's aphorism, "Think globally, act locally." I think that it is some of the best advice I have ever heard. I never come to a session like this without learning something that can be used in the local decision-making process, the daily grind we go through trying to get decisions made the right way. How do you feel about this?

ANSWER. There is also the question of how to work from the local political base for changes on the local level. I think that going in directions different from this is always a danger. There is a lot of valuable information at the grass roots level that should be working its way up to the global level.

II Conservation Objectives

Ecological Processes and Life Support Systems

GERARD A. BERTRAND

Most people in the United States and developed countries are familiar with efforts to save endangered species. Large, dangerous, or spectacular animals or beautiful rare ones instantly attract attention. But the plight of endangered species is only the most obvious symptom of a much more widespread disease involving the deterioration of life support systems on much of the globe.

The World Conservation Strategy (IUCN, 1980) recognizes immediately that if the international objectives of conservation are to be achieved there must be a fundamental understanding of the earth's ecological processes and life support systems and both citizens and governments must recognize that these processes and systems are the basis of life itself. The strategy raises three key areas for specific attention: the maintenance of our worldwide agricultural systems and the soils on which they exist; the maintenance of forests as prime watershed protectors and the key to retention and formation of soils; and the maintenance of coastal and freshwater systems. All are critically important to biological production. It is obvious that these choices were made in developing the brief strategy list because the fundamental ecological processes critical to earth involve land, water, air, and living things as well as subsets of them.

This chapter discusses aspects of each of these areas and some of their connecting links. These links are perhaps more clearly seen in discussing ecological processes than in any other specific area covered by the strategy—the mixing of physical and biological processes on or in the land and air is the basis for life itself. Without these processes the dominant species in each ecosystem, including man, could not exist.

Topsoil and Agricultural Systems

The *Global 2000 Report to the President* contained a series of projections of growth rate until the year 2000 (CEQ, 1980). From the earth's 4.1 billion people in 1975 the median projection indicated a 6.35 billion world population in the year 2000. More recent projections have indicated 6.1 billion for the year 2000 (Winder, 1982). Of this population growth, 92% will occur in developing countries. Human population growth is the fundamental driving force for imbalancing our ecological processes and systems. This growth without planning or care can lead to disaster. Malthus was right, only his timing was wrong.

No fundamental ecological process has been more disrupted by population growth and its demand for food than soil building. Because we are land animals we forget that 70% of the surface of the globe is covered by ocean. We live in a water world. Of the 30% of the earth's surface that is either land or fresh water, only 11% has the capability to grow food without additional energy inputs from irrigation, drainage, or fertilization (IUCN, 1980). Since the World Conservation Strategy was written, that percentage has dropped to 10.5. Because the aim of this book is to bring the World Conservation Strategy down to a regional level, it is worthwhile to examine what has happened with topsoil in the United States.

The United States has many strengths, but chief among them are its natural resources of land and clean, fresh water. The largest single productive area of arable farmland on earth lies in North America. Over 400 million acres of the 2.2 billion acres in the United States is farmland (Sampson, 1981). To put this in perspective, Massachusetts in toto is only 5 million acres, one-eighteenth of the total farmland in the United States. Until the 1970s much of this farmland had not been cultivated, or at least not cultivated since the dust bowl disaster of the 1930s. With more than one-third of the

world's best arable farmland, the United States is in the strategic position to produce food, and it has the responsibility to protect that production potential. It is, in fact, the most important food growing area on earth.

If the present population growth rates continued and the rate of food production per acre stayed stable, we and other farmers around the world would have to increase the amount of arable farmland by one-third. Without new land we must increase production per acre.

What has happened to American farmland with increased production? The dust bowl made soil conservation a part of U.S. national policy. In 1935 President Roosevelt signed the Soil Conservation Act into law saying,

It is hereby recognized that the wastage of soil and moisture resources on farm, grazing and forest land of the nation resulting from soil erosion is a menace to the national welfare and that it is hereby declared to be the policy of the Congress to provide permanently for the control and the prevention of soil erosion and hereby to preserve natural resources, control floods, prevent impairment of reservoirs, and maintain the navigable rivers and harbors, protect public health and public lands, and relieve unemployment (U.S. Congress, 1935).

This proclamation and law did not stop the enormous soil erosion occurring in the American Midwest, however. Hugh Bennett, in a report to Congress in 1939, said that each day the equivalent of twenty-four Missouri farms were still being eroded away and that close to 300 million acres of land had already been lost in the United States since farming began (1939).

The loss of American farmland really began in earnest in 1833, when John Deere invented the millboard plow—a powerful tool that was capable of digging deep into the western soil and thus making weed control possible (Morganthau et al., 1982). The deep straight furrows created by this plow became symbols of excellence in agriculture the world round, and American farming methods were regarded as the ultimate. But till plowing led to erosion, particularly when little regard was paid to other necessary land uses. The disaster of the dust bowl became so great that it inculcated lessons into American farmers that survived through the

FIG. 3.1. *Terrace agriculture on former mountain forestlands in Nepal. While terracing is often a sound farming practice, pressure from a growing population has led to its use on increasingly unstable slopes. Photograph by author.*

1960s. These lessons included contour plowing, hedgerow planting, terraced farming, planting to avoid swales and wetlands, crop rotation and the practice of letting land lie fallow for a year or two on a rotating basis to renew soil nutrients.

From 1950 to 1970 the United States had increased farm production by almost 50% (Sampson, 1981). The greatest problem for the U.S. government during this period was how to manage surplus farm products. Many still remember when the government had so much excess production that it was dumping butter into the sea and practically giving away hundreds of millions of tons of wheat to developing nations in Asia and Africa. Piglets were being killed to avoid the problem of surplus pork. All this production required the input of nutrients, pesticides, and herbicides. The use of these new inputs necessitated more expensive and sophisticated equipment. This, in turn, led to bigger, more technological farms and fewer farmers. Much American agriculture is presently held and controlled by absentee landowners (Burger, 1978). In addition,

there is less stability in this farming method, because high yields from nitrogen input are more weather dependent.

A rude awakening came in 1972 when, without the foreknowledge of the U.S. government, the Soviet Union bought 18 million tons of grain, effectively driving up U.S. prices and wiping out the usual American surpluses. This windfall for farmers was welcomed, and the administration saw in farm exports a new way to balance fast-rising oil import costs (by 1975 oil had gone from $2 to $35 a barrel). Earl Butz, then secretary of agriculture, told farmers to plant "fencerow to fencerow." The resulting farm production sent U.S. farm exports from $7 billion in 1970 to $35 billion in 1979. In exhorting farmers to grow more, the government turned its back on thirty-five years of soil conservation. This, in conjunction with agribusinesses' assault on the land by the sheer size and character of their operations, turned back the clock on soil conservation. Hedgerows were cut out, swales were filled, and wetlands were drained. Remember the tractors on the Mall in Washington, D.C., as part of the farmers' demonstrations during President Carter's first year in office? Many were air-conditioned tractors in which the farmer was well removed from the soil.

The economic problem, of course, was that the stimulation of production by farm exports eventually led to lower domestic prices. Farmers could not keep up with the costs of inputs to farming, particularly the costs of energy. They were using more energy, fertilizers, and pesticides and working harder than ever, but making less for it.

The important ecological lesson to all of us is that what happened with soil loss in the dust bowl is now happening again. Not as spectacularly, to be sure—there are no dust clouds blowing one thousand miles out of the Midwest to New York and Boston. But the topsoil, that unique resource which keeps us all alive, is eroding away at an alarming rate. Neil Sampson, in his book *Farmland or Wasteland: A Time to Choose*, estimates that almost 4 million tons of topsoil are lost per year in the United States (1981). At this rate 6 inches of topsoil from arable farmland will be gone in the next one hundred years—a 20% loss of soil for us and the world community by the year 2000. This is a loss to the whole world, because it means millions fewer people can be fed. And it is, for all practical purposes, irretrievable.

Topsoil formation is an ecological process—a fundamental one—which even at its best is very slow. In the savanna grasslands in Africa it takes 1,000 years to deposit 1 inch of new topsoil. That same soil in a tropical rain forest might take 10,000 years to form (IUCN, 1980). In tropical forests nutrients are tied up in the forests themselves. Annual production adds to the bulk of the forests and not directly to the soil. There is no simple, economic way to build topsoil quickly. Topsoil is a slow accumulation of the excess of material produced from solar energy minus losses to a wide variety of environmental factors, including biological ones. Of course, the earth itself can add to soil building. The Mount Saint Helens eruption added spectacularly.

A major deposition of nutrients from that explosion dramatically increased soil fertility and production in Washington, Oregon, and Idaho. Ashfall doesn't happen very often by human standards, but geologically it can be significant.

The slash and burn technique of agriculture temporarily adds to the topsoil. This practice consists of cutting, drying, and burning forests to release the nutrients and return them to the soil. The resulting temporary enrichment may give a year or two of harvestable crops in tropical areas. But once the crops and their nutrients are removed, the remaining soil is nutrient depleted, subject to erosion, and a poor base for future forests or agriculture. Most tropical soils are thin, lacking the glacial depositions that so enrich temperate regions. If a slashed and burned area is small enough, the soil can, with time, be enriched by debris, leaves, branches, falling trees, and other materials from surrounding forests. But if large areas are stripped, the topsoil can be permanently destroyed. The slash and burn process is one of the major reasons topsoil is being lost at such an alarming rate. Fifty percent of the soils traditionally tilled in mountainous areas of Nepal are already badly eroded.

There are now political disputes between Bangladesh and India over the newly forming islands in the Bay of Bengal. In reality, however, these islands, which create deltas hundreds of square miles in extent, really should be claimed by Nepal, because it is their topsoil washing from the Himalaya Mountains that has been entering the bay. The floodplain's size has been increased four times in the last few decades. In a real sense, the best of Nepal's future—the fertility capable of protecting living things, including people—is now in the ocean. Nepal is a spectacular example, but the United States, which has far more topsoil to lose, is losing it at rates that are just as alarming. In Iowa one hundred years ago, the average topsoil depth was 16 inches—the present depth is only 8 inches. In Illinois 181 million tons of topsoil are estimated to wash away every year—that is 2 bushels of topsoil for every bushel of corn produced.

Besides direct loss of topsoil, we are losing the fundamental life support of farmland in another way: through conversion to other uses. Every hour 220 acres of farmland are converted to town houses, industrial centers, family homes, or shopping malls (Sampson, 1981). This is the real mauling of America. Each day this loss is equivalent to twenty-three average-sized Missouri farms. New England in the last hundred years has lost one-half of its farmland to conversion. Massachusetts itself has lost more than 50% since 1940. The Mid-Atlantic states have lost a quarter of their farmland and the Midwest one-tenth (Morganthau et al., 1982).

The pressure for conversion continues unabated. Without statewide or nationwide land use protection, there continues to be a patchwork of development, which maximizes short-term economic uses of land—usually for development—rather than long term production. Too many farmers have been lured into selling their land for development so they can retire to Florida. Too much farmland has been held for speculation by absentee institutions such as banks, insurance firms, or energy companies with little regard for farm values. The economic system, including our taxation system, favors development over protection. President Roosevelt said: "The nation that destroys its topsoil destroys itself" (Roosevelt, 1937). We in America have taken our bounty of topsoil and farmland so much for granted that we may end up denying our grandchildren. Pope John Paul II put it more eloquently in Iowa in 1979 when he said: "The land must be conserved with care since it is intended to be fruitful from generation unto generation. You are stewards of some of the most important resources God has given the world. Therefore, conserve the land well so that your children's children and generations after them inherit and leave an even richer land than was entrusted to you" (Reeves, 1980).

Erik Eckholm, in his examination of agriculture and topsoil protection around the world (1976), documented the international nature of the soil erosion problem. It is even more frightening to realize that these losses have been going on for millennia. The Middle East may once have been one of the more productive farm areas on earth. The Fertile Crescent, Mesopotamia, the Iranian heartland, and the rich north coast of Africa must have had sufficient topsoil to support their well-developed civilizations—but that topsoil is gone, blown away after years of debilitating overgrazing and perhaps overtilling. Now much of the area is desert. I have seen the same thing happening in Pakistan and India. Much of the American rangeland suffering from overgrazing is, itself, subject to windblown erosion.

There is no question that maintenance of our food production systems requires the protection of our topsoil through farming methods that encourage soil maintenance and development.

Besides the loss of farmland itself I should mention one other aspect of our agricultural systems under threat. We need to maintain crop pollinators. These beneficial insects, so significant to productive agriculture, have been the victims of the ever-expanding chemical war on agricultural pests. The reduced population of bees is a limiting factor on production in some areas of the United States. Wild populations of natural pollinators, including moths, bees, hummingbirds, bats, and other nocturnal creatures, can exist only if a land use pattern that mixes forest and natural vegetation with agricultural lands is maintained. The chemical warfare waged on agricultural pests, like the chemical enrichment of the soil to feed plants is expensive, energy consumptive, and detrimental to the natural ecological process that allows long-term agricultural production.

Forests and Forestation

Few people think of forestation as a fundamental ecological process. But in fact there is a constant battle between the physical elements and wind, water, and fire to determine whether grasslands or forests possess the surface of the earth. Maintenance of the earth's essential forest cover is important for a wide variety of reasons, including water retention, soil protection, climate modification, and oxygen production. In addition, however, forests produce products that are needed by many of the world's people. Sixty percent of the earth's population rely directly on forests for their fuel needs. Most of these people are in the less developed countries. For example, out of 780 million people in India, close to 400 million use firewood and another 300 million use cow dung as a principal fuel source. In developed countries forest products are used far less for fuel than for paper and construction materials. "Each year the average American consumes about as much wood in the form of paper as the average resident in many Third World countries burns as cooking fuel" (Eckholm, 1979).

Because forestation is a process, human activities can affect both its rate and the extent to which it occurs. Fire control in the United States has greatly affected both the extent and character of American forests. Land use changes in New England have taken place so rapidly in the last hundred years that more than 6 million acres of former pastureland and farms have returned to forest. In viewing the forests of the world as a life support system, there are great differences between the forests of the far north—the coniferous biome, which as a whole is in reasonably good condition and still extensive—and those in the tropics. Coniferous forests and forest cover in Scandinavia, the Soviet Union, and North America are stable or expanding. In developing countries on the other hand, particularly tropical ones, there has been a rapid disappearance of both moist and dry tropical forests. The extensive dry tropical forests disappeared first, because the land on which they existed was most easily convertible to farms, ranches, towns, and cities. Some scientists consider dry tropical forests to be the most endangered forest biome in existence, but the diversity of living species and the complexity of these forests is far below those of the moist tropical forests, which may hold 50% of all the earth's species (Myers, 1979).

Although tropical forests are among the oldest, most magnificent, and most diverse of the forests on earth, they comprise a small portion of the total forested area—less than 7% of the earth's surface. These forests have a disproportionately important effect on the world ecosystem because of the sheer height and biomass of their forest cover and because of their complexity and diversity.

The *Global 2000 Report to the President* (CEQ and Department of State, 1980) Norman Myers's *The Sinking Ark* (1979), and numerous other recent works have estimated that 15% to 20% of the world's fauna will disappear by the year 2000. Ninety percent of this loss is expected to come in the tropical forests. The loss of tropical forests can also have significant effects on local and regional

FIG. 3.2. *Goat herders cutting greenery from one of the few remaining trees outside a village in Rajasthan, India. Photograph by author.*

climates. The water retention capacity of forests is of major importance in preventing floods. Deforestation followed by rapid siltation behind dams and reservoirs is a disastrous phenomenon for developing countries, which need both maximum reservoir capacity and the hydroelectric power generated by dams. Without forests, the monsoon cycle in Southeast Asia becomes a devastating force for erosion, rather than a life-giving force for renewal. Once a country enters the flood-drought cycle, which occurs in vastly deforested areas in the tropics, there is little hope of natural renewal without massive artificial planting and management (Eckholm, 1979).

The most extensive areas of protected forests in the tropics currently exist in the African Congo, the Amazon River basin, and the island forests of Southeast Asia (IUCN, 1977). It is precisely these areas that will undergo the greatest deforestation in the coming two decades. The Global 2000 study's technical report indicates that the growing stocks of commercial-size wood in less developed countries will decline 40% by the year 2000 (CEQ, 1980). Tropical forests are currently being destroyed at rates between 600 and 700 square kilometers per day. To put this in perspective, an area about the size of Massachusetts is clear cut each month. India's natural cover at the turn of the century was approximately 33%. Official government estimates put present forest cover at just over 20%, but conservation organizations and satellite photography indicate that its true value may be no more than 12%, at least 4% of which is artificial forest cover, such as eucalyptus or teak plantations.

Forests are a fundamental part of the life support system on earth. I know of no estimates of the minimum forests necessary to maintain water and soil values and the genetic systems on which so much of our agricultural and pharmaceutical industries depend—nor those needed directly for human fuel and shelter. But we do know that forest cutting has already progressed to the point that in some areas irreversible damage is already occurring. Historic precedents for this abound. The Cedars of Lebanon, so justifiably famous 2,500 years ago, were in fact major forested areas. Historical records indicate that the northern fringe of Africa was forested—dry forests to be sure, but they supported an excellent fauna. These forests are gone. Intensive grazing and cutting were followed by a topsoil loss that led to irreversible degradation.

Desertification, one of the results of too-rapid deforestation, is occurring in Asia, Africa, Central and South America, and even the United States. The World Conservation Strategy states that desertification is a response to the inherent vulnerability of the land and the pressure of human activities. Regions already in the grip of desertification or with high to very high risk cover 20 million square kilometers—an area twice the size of Canada (IUCN, 1980).

The only hope to maintain our forest resources is a dramatic reversal in present cutting trends in the Third World. A solar cooker replacing the constant

FIG. 3.3. *Indian woman carrying firewood from the buffer zone of Corbett National Park, India. Photograph by author.*

need for firewood of more than 2 billion people in the tropics would go far to lessen pressure on the forests. Slash and burn agriculture, which in areas like Thailand has led to the disappearance of more than half the forest growth and resultant soil loss, can only be discouraged by teaching farming practices that allow renewability of the forest and maintenance of the topsoil. Only governments are capable of effecting such a massive change in attitudes and present practices. Similarly, major efforts are needed to replant forests. The Indian government has taken this on as a serious task in the last three years. In India billboards saying "Save the Country, Plant a Tree" are becoming more obvious thanks to international and national programs that make the relationship among forests, agricultural production, and drought publicly recognized. This relationship is already well appreciated by the upper echelons of Indian society, and their awareness is beginning to spread toward the agricultural sector as well.

There is not yet an accepted international plan of action for preservation of tropical forests. The United States, which should be a leader in such efforts, and which was going in that direction under the Carter administration, has dramatically slowed its work for tropical forest protection. Laissez-faire economics and long-term forest protection have little in common, because capital-starved countries can turn forests into quick profits, which can be used to solve immediate problems for a short time. The U.N. Food and Agriculture Organization is the most obvious focal point for an international preservation program, but to be effective it needs strong national support.

Water

Just as topsoil is being mined around the world by constant plowing and erosion, the world water cycle is being supplemented by mining groundwater in a nonrenewable way. "Primal water" formed eons past in many areas is being brought to the surface and used. Renewal of near-surface groundwater is a physical process complemented by grasslands and forests, which act to retain topsoil and minimize water losses.

With increasing population, declining quality of the soil, and decreasing available arable lands, irrigation will increase dramatically by the year 2000. In fact, a doubling of water use for irrigation is expected between 1967 and 2000 (CEQ, 1980). Intensive irrigation is often accompanied by the salinization of soils and an eventual loss in productivity. As serious as the energy shortage of the mid-1970s was for many developed countries, the water shortage of the future will be more widespread and affect more lives. The loss of water is expected to be accelerated by deforestation in the tropics. Without forests thin tropical soils become dry hardpan.

Besides the increasing use of water for human agricultural and industrial needs, water will become less available because of pollution and toxicity. The essentially free nature of our past water supplies is no more dramatically illustrated than when a town must stop using its own sources and purchase new water to replace a contaminated system.

Increasing acidity from sulfur and nitrogen oxides in the air has led to deterioration of fresh waters and their biological productivity throughout Scandinavia, and now the same effects are becoming dramatic in the American Northeast and eastern Canada. The clear cause and effect relationship between industrial discharges and increases in the acidity of surface waters has been too well documented to ignore.

Another local threat to water supplies is the increased use of road salt and subsequent salinization of water supplies. In Massachusetts more than a dozen towns have such elevated salinity levels that their water cannot be consumed by heart patients or others on low-salt diets. In some cases salinity is so high that water supplies have been contaminated for all citizens.

The ironic thing about the water problem is that countries with rich water supplies tend to take them for granted and waste them lavishly until costs rise and supplies are imperiled. We have not taken water conservation seriously. Americans are used to government-subsidized water, so we have become accustomed to abuses such as using 3.5–5 gallons of water every time we flush the toilet. Whether in drought-torn Arizona or water-rich Michigan, we still think nothing of watering our lawns in the middle of the day when evaporation is at its maximum or leaving the tap running while we brush our teeth or failing to install low-flow shower heads or aerators on our faucets. These are measures that would save enormous amounts of water, sometimes up to 50% of expected household use.

If we were more like those in the Third World, who have to carry their water several miles, no water would be wasted casually. With ever-intensifying pressure on water supplies and increasing contamination, there is bound to be an increase in waterborne diseases and parasitism. Many people are surprised that waterborne diseases can reach some of their highest incidences in desert areas. The Egyptian epidemic of schistosomiasis described by William Stapp ("Building Support for Conservation and Education" later in this book) is a good example of this phenomenon.

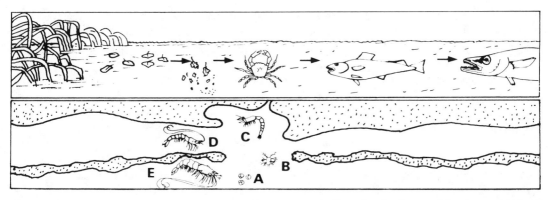

FIG. 3.4. *Coastal wetlands, essential to the productivity of marine fisheries, are a primary focus of international conservation attention. The top illustration shows the flow of nutrients from a mangrove swamp to offshore fisheries. The drawing below shows life cycle stages of commercially valuable shrimp: (A) the eggs are laid offshore; (B) the larvae move inshore; (C) juveniles and (D) adolescents then shelter and feed in mangrove swamps, estuaries, and lagoons; (E) the adult shrimp return to sea. Illustration from IUCN, 1980.*

Water protection is to a very large extent dependent on land protection. Floodplain and wetland protection, identification and protection of aquifer recharge areas, and forest protection are all parts of the process of maintaining our water supplies.

Coastal and Freshwater Systems

The World Conservation Strategy places heavy emphasis on maintenance of coastal wetlands and shallows as well as freshwater areas, largely because of their critical importance to the production of the world's fisheries. Shrimps, scallops, crabs, and many commercial fishes need natural coastlines for protection during at least some part of their life cycle. Approximately 70% of the world's fisheries have some direct relationship to these coastal zones.

Coastal shallows are particularly rich in production because of the downstream flow of fresh water and nutrients to them. The ecological process of enrichment by riverflow is critical to this production (Bader, Rogotzkie, and Teal, 1972). While dams and upstream activities have decreased the downstream flow of nutrients in some parts of the world, particularly the American West Coast, they have increased the flow in others, where heavy erosion and forest cutting have been uncontrolled. In some areas there is a major loss of productive fisheries from coastal degradation. The strategy estimates $86 million in lost fisheries per year in the

United States (IUCN, 1980). I cannot help feeling that this figure is far too low. In the Ipswich River of Massachusetts alone, it has been shown that close to $500,000 worth of soft-shell clams are not harvested because of bacterial contamination from septic systems associated with summer homes (Hruby, 1981). This bacterial contamination is the rule rather than the exception in many developed coastlines in the Northeast.

In spite of existing degradation, however, coastal pollution generally has declined in the United States. Other countries, too, have taken major steps in cleaning their coastal waters. Singapore is an outstanding example of what can be done in a crowded developing nation. The recurrent problems of the Mediterranean are at last being addressed and the steady decline of its coastal waters seems to have been halted, but too many countries around the world still have little or no formal protection system. Coastal development and resultant waste disposal continue to be the two greatest problems.

Although the coastal zone—taken as a whole on the world scale—is in moderately good condition and many of its changes are reversible, the land-sea area continues to be under the highest pressure. It is in the coastal zone that the demands for recreation, power plant sites, marinas, summer homes, commercial and sport fisheries, industry, harbors, and military uses all converge. Given a choice, most people want to live near the ocean. Historically, development of shoreline areas occurred because of access to commercial trade and past exploration patterns. Present development seems to be reinforced by a preference to be near the water for aesthetic and recreational reasons. This concentration of people and their needs can lead to disruption of coastal zone ecological processes. There is no question that pressure will continue to grow as new industries, such as mining of the continental shelf, are developed. Currently, most coastal mining is restricted to gas and oil development and mining of sand as a construction material. Future mining development, particularly of the continental slope and ocean bed, could have serious consequences in the coastal zone—particularly from the ancillary development of the coastline as a staging, storage, and transport area for that mining. The refusal of the United States to sign the new Law of the Sea Treaty is broadly viewed with concern and could have serious impacts on future coastal zone protection around the world.

A specialized form of coastal mining for calcium carbonate has been and is occurring on many coral reefs. This was brought home clearly to me when I was investigating a proposed new national park in southern India and watched sixty people dynamiting coral reefs on the border of the proposed park to extract calcium carbonate. Although this sort of abuse to coral reefs is relatively unusual, much less traumatic disturbances to the reefs can also have major effects. Reefs are fragile communities, and they are subject to degradation from coastal pollution or destruction from the sedimentation associated with coastal mining. They are vulnerable to almost any change in water quality or clarity.

The World Conservation Strategy gives particular emphasis to the coastal zone because of its enormous value for feeding a portion of the world's population. The world fish yield presently seems to have stabilized around 70 million metric tons per year. Sixty million of these come from the marine environment and the additional 10 million from

fresh waters. These yields have been essentially stable since the mid-1960s, but they could increase dramatically if krill were exploited to any extent. Estimates of krill yield vary from 50 million to 200 million metric tons per year. Such harvests, however, are probably at least a decade away.

The freshwater systems, which account for about 14% of the world's fishery production, have decreased greatly in the last hundred years. For example, the United States has lost approximately 50% of its freshwater wetlands and with them the production of ducks, geese, and other species that utilize those areas. The Massachusetts Wetland Act (MGLA ch 131, section 40) was the first in the nation and remains one of its strongest. It came too late, however, for most Massachusetts wetlands, because significantly more than half had disappeared before 1970. The worldwide recognition of wetlands as productive and valuable areas has been one of the most gratifying results of the environmental movement. In spite of this, however, the International Convention on Wetlands of International Importance, Especially as Waterfowl Habitats (1972), remains unsigned by most Western Hemisphere countries, including the United States. At present there is no guaranteed international protection of major world wetlands. The issue is still looking for an international champion.

Ecological Processes and an Overall Perspective

The last major area I wish to discuss is ecological processes. There is a balance among carnivores, herbivores, and the plant systems in our environment. Each change in the environment causes a shift in that balance, which is important to us and to our survival. For example, 60% to 70% of all medicines we use are extracted from natural products of plants and animals (Myers, 1979). I have already mentioned the importance of pollination of crops by wild species of insects, birds, and bats. We rely on natural systems for our own survival and as part of our life pattern from day to day.

We are at present dramatically changing the composition and variety of living things on earth. Many species, once abundant, are threatened or endangered. These species, although not biologically extinct, are ecologically extinct—they no longer are major influences on the composition of animal or plant communities. The difference in the composition of the Antarctic Ocean fauna caused by the reduction of the blue whale population from 1 million to the 10,000 that may now exist may well be critically important to that ecosystem and to the whole globe. Besides changing animal and plant composition through habitat modification or purposeful elimination of individual species, we dramatically alter biological communities by the introduction of new species. This manipulation has progressed to such a state that few if any areas of the globe are free of exotic plants and animals.

Fire is an essential ecological process that has been largely restricted in most parts of the world. We cannot afford, from an economic standpoint, to let lightning fires run across the American prairie as they probably did, nor let savanna fires burn themselves out in Africa or Asia. Consequently, fire as a natural process has been much controlled, and this action has altered natural systems.

We probably now have the technical ability to modify drastically weather patterns, particularly hurricanes, and lessen the effect of winds and rain on our coastal areas. In Florida, where I spent three years working on water problems, it became very apparent to me that without hurricanes that state would miss the major influx of water needed to recharge aquifers and halt the advance of salt water and salt vegetation. The same technical ability to modify weather may eventually allow us to control monsoon rains in the Asian tropics. Perhaps of far more worry, however, than the ability to alter climate purposefully is the inadvertent climate modification associated with changing land uses and the disappearance of forests. Other processes, such as reducing atmospheric ozone and increasing carbon dioxide in the world atmosphere, are almost too large to contemplate. They could interfere with both the extent and distribution of polar ice caps and the formation of Antarctic Ocean waters, which feed the deep areas of the entire globe.

Physical processes and biotic ones are interrelated in ways we neither understand nor can predict. The difficulty, of course, is that all environmental topics are interrelated. Farmland and topsoil protection is part of the overall problem of protection of the land. The national recognition that is needed to allow us to live and develop land without interfering in the fundamental ecological processes that support us in most cases does not yet exist. Perhaps we need a statewide zoning, such as that in Oregon, or even a national zoning that recognizes the overriding importance of topsoil and farmland, wetlands, and aquifer recharge area protection. National zoning is not completely out of the realm of possibility. Switzerland and Denmark have approached it already, and their systems have been largely successful. Land has been treated as an unrestricted commodity for so long that there is a deep resistance in many parts of our country to such federal or state control. The speed at which we overcome this resistance and recognize our own ecological needs is linked to both the population and resource pressure we will receive from around the world.

Maintenance of essential ecological processes is not a job that private citizens or private organizations, no matter how well intentioned or well endowed, can accomplish on their own. It requires a combination of private, national, and international action to be successful. The present worldwide economic slump has hindered governmental and international abilities to address problems with ecological processes, but it has also slowed rates of development, building, and population growth. There is a growing understanding, particularly in Third World countries, of the interrelationship of land, water, forests, and the economic future. The planning necessary to awaken all governments to these needs has been happening for almost a decade. The World Conservation Strategy is only one evidence of that awakening. But to maintain the ecological processes essential for life and slow the degradation in these processes will take more than just awareness. It will take a fundamental change in life-styles and resource use. We cannot save topsoil in America without fundamental change in farming systems, nor can we conserve water without adjustments in both how it is costed and how it is allocated. These efforts—and the overriding one of reducing population growth rates—are the most fundamentally important challenges facing mankind.

Discussion

QUESTION. Much of Central America is clearing forest to raise cattle, which will help to meet the demand for beef in the United States. Is this deforestation severe enough that more attention should be paid to it?

ANSWER. I do not have personal experience in all of Central America, but I have seen what deforestation has done in some parts of southern Mexico, where U.S. industries have bought large pieces of land. The first thing they do is move the campesinos off the land. These people had no more than a subsistence life-style to begin with; they grew a little maize and raised a few chickens. The U.S. ranching interests fertilize, irrigate, and put in buffalo grass, which is totally unproductive for wildlife. They displace indigenous people and ruin an ecosystem to grow cheap beef for American markets. It is an international rip-off. We are using Mexicans to subsidize the American life-style. This may not be biologically significant on a global scale, but it has profoundly affected local areas.

The policies that the Japanese, German, and American lumber industries followed until recently may be more significant. Tropical forests were destroyed to bring mahogany to Europe and the United States so that we could panel our walls. We had tropical hardwoods coming to us at very low costs, while the producing countries suffered enormous losses of water and topsoil. There were virtually no environmental controls—if you paid the right price you could do nearly anything. Our luxury is being subsidized by the future environment of other countries, usually without the knowledge or support of anyone except a few political officials.

In just the same way, the United States is exploited as an undeveloped country. Our redwood forests are being exported to make decks in Japan. The Japanese are not cutting the forests they have. We are on our way to becoming an export economy. We are busy shipping our natural resources overseas, sending them to the rest of the world faster than we can replace them. We are not a value-added country. We do not process our resources and then send manufactured goods overseas the way the Swiss and the Japanese do. Everything that goes out of their countries is a manufactured product, a value-added product. In spite of the fact that they are bringing in oil worth billions of dollars, they still have an excess balance of payments.

No country can make it in the long run if it simply ships its resources out as fast as possible.

QUESTION. The World Conservation Strategy implies that the solutions to the problems it documents demand dependence on centralized public institutions. They are supposed to command and regulate. In the last few years the federal government's responsiveness to the environmental movement has been shrinking. Do you see conservation shifting from reliance on political institutions to the private sector as governments fail to show real leadership? Who is going to fight for the solutions the strategy demands?

ANSWER. There is going to have to be more reliance on nonprofit organizations, especially local ones. Certainly, the environmental movement is absolutely delighted to work with a public official who is "one of our own." But in the long run, this may not be the most effective way to get the work done. We may need to become much more autonomous. Studies at Yale showed that 70% of the American public strongly supports environmental controls and will pay from their own pockets to support them. We have to remind the politicians that they ignore environmental issues at their own risk. The public has not backed off, even if they have.

I also think we have to get much closer to the business community. We both have the same goals in many ways. The business community applauds almost everything the Massachusetts Audubon Society does. When they withhold their support, it is because of the 1% of our activities that they dislike. We must build a coalition to fight for the long-term future. There will be times when environmentalists and business leaders argue, but that does not mean we cannot talk and support each other.

Environmentalists also need to develop stronger ties to people working for population control. It is inevitable that a bigger population means less individual freedom. You cannot defeat this correlation anywhere. Centralization inevitably creeps in, and with it less chance to affect the environmental future that the government has planned. Compare China to India. Together they are 37% of the globe's population. China is very well organized centrally. It has to be. India is really fourteen countries thrown together. There is continual civil war—chaos. The only way India will be saved is to be centralized. In the United States, our grandchildren will not have the same freedoms that we have if there are 600 million of them.

Environmentalists are going to have to develop their own political power. Part of it will come from the coalition building I have just described, but the rest must come from within. We cannot rely on the central political machinery alone to implement our programs.

QUESTION. I am sure that there are institutional differences among countries that make environmental controls superficially different, but are there social or psychological differences as well? Do they interfere with decision making at the international level?

ANSWER. America is a very loose country. We try something out, if it doesn't work, we change it. This does not bother us. It does bother most Europeans. They prefer things that are stable and settled. Every country has its own patterns. When the Convention on Trade in Endangered Species was being drafted six or seven years ago, the Americans and the British spent much of their time arguing. The British kept telling us that the convention was unenforceable, that no one could be sure it would work every time. We said 90% of those concerned would obey it voluntarily. These are two totally different approaches to law. They are typical of the difference between European and American outlooks.

Another example is our environmental impact system. Basically, it works, but it has a long way to go. There are still too many delays, wasted dollars, and regulations. Europeans want a much neater package. Many countries have adopted environmental processes very different from ours. Most would be very unhappy with our system as it stands. The Japanese, for example, would be aghast at how little room there is for public discussion of our governmental decisions. Still, cultural variety is usually not an insurmountable difficulty. In fact, it is one of the great sources of invention at the international level.

QUESTION. What do you see as the biggest obstacles to environmental protection in developing countries? Will the developed countries be able to give them any real help?

ANSWER. Developing countries need education to understand the issues, a political infrastructure to manage the issues, and, finally, capital. All of these are absolutely essential but enormously difficult to provide. However, Third World countries are becoming more and more aware of the need to maintain their own environmental quality by watching our mistakes. Indira Gandhi says, "Plant trees, because if we do not plant them, no one else is going to." In Africa they realize that if they do not keep their wildlife, tourism—which is so important—will disappear. They have watched it happen in other countries. The Third World countries will build their own futures. We can assist them, but the real drive must come from their own grass roots. If it does not, no outside help will have effect. We have tried to impose Western values and solutions all over the world, and we have failed in most places. They do not fit the cultures, life-styles, or desires of most developing countries.

It is often more useful to export an idea. The environmental impact statement process is a good example. It has been adopted all over the world, by less developed as well as developed countries; more than seventy-five nations now require environmental impact assessment. When India wanted to build the Theri Dam, it set up an interagency environmental committee. Judges in the national courts required assessments of the dam's effect on native people, wildlife, long-term production, and many other factors.

Another idea that has been exported successfully is maximum sustainable yield (MSY) production—the notion that you should not take more of a resource now if it means taking less later. This idea went all over the world. However, MSY does not take into account the complexities of the natural system. It also misses the social factors that are often part of a society's relationship with a resource such as wildlife. So now we have an improved export that has been accepted just as readily—MSY plus a safety factor.

If we understand our own systems and our own mistakes, our ideas will be our best exports.

QUESTION. Do you feel that the Third World is developing conservation *skills* in time to preserve enough of its diminishing base of species, habitats, and resources?

ANSWER. Some countries are not. I have very little hope for Pakistan or Bangladesh. On the other hand, Honduras, Costa Rica, Panama, and Ecuador can probably meet the challenges they face. Success with environmental problems usually depends on the ability of a country to carry over ideas from one government administration to the next. Pragmatically, stability determines how much capital will flow into a country from the rest of the world. A country that reverses its policies regularly simply will not attract investment—and the problems are compounded if the country has a high birthrate. India is getting closer to stabilizing because it is finally controlling both political upheaval and population growth.

In a different sense, the Third World is our hope for developing conservation skills. They have to reform us. They have to help us mend our extravagant and wasteful ways. Unfortunately, at present their own population growth is what is putting the most pressure on us to reform. No country is in a position to teach conservation skills on the basis of its own unqualified success.

Preservation
of Genetic
Diversity

F. WAYNE KING

Preservation of wild species and the genetic diversity they represent is possibly the single most important task facing mankind. No one reading these words would be alive today were it not for wild species and their domestic offspring. The meals you ate today undoubtedly were derived from wild species. Some article of your clothing probably was produced from wild species. Wild species were probably the basis for any medicines you took today. Even the paper this book is printed on was derived from wild species. Mankind depends on them in a very real way. Many of our necessities, as well as our luxuries, come either directly from wild species or from the livestock and crops that are their offspring.

In the future, benefits will be obtained from wild species that have not yet even been discovered. Unfortunately, governmental officials and a large segment of the general public seem unaware of their value. Too few seem concerned that wild organisms are becoming extinct at an ever-increasing rate and that once extinct they are irretrievably lost, to man's detriment.

Professor E. O. Wilson of Harvard University outlined the problem in a rhetorical question:

What event likely to occur in the 1980's will our descendants most regret, even those living a thousand years from now? . . . The worst thing that can happen—will happen—is not energy depletion, economic collapse, limited nuclear war, or conquest by a totalitarian government. As terrible as these catastrophes would be for us, they can be repaired within a few generations. The one process ongoing in the 1980's that will take millions of years to correct is the loss of genetic and species diversity by the destruction of natural habitats. This is the folly our descendants are least likely to forgive us (1980).

Extinction and Diversity

The fossil record shows that before the appearance of man, a species became extinct about every 1,000 years (Myers, 1979). At that rate, evolution could produce species faster than they disappeared. Over time species generally became more numerous (Ehrlich and Ehrlich, 1981). However, when humanity arrived on the scene things began to change. Man's activities, particularly the spread of agriculture and advanced technologies, accelerated the rate of extinction by modifying habitats and overexploiting some species. Because of man the rate of extinction now far surpasses the rate at which species are created. History records that during the 1400s and 1500s man caused the extinction of the tarpan (ancestor of the modern horse), the aurox (forest ox progenitor of domestic cattle), and a few other vertebrate species (Allen, 1942; Greenway, 1967; Harper, 1945; Honegger, 1981). During the 1600s and 1700s the number of species that man consigned to oblivion grew to several dozen. Included among those were the dodos, the large flightless pigeons that lived on the Indian Ocean islands of Mauritius, Réunion, and Rodrigues (Greenway, 1967).

During the 1800s the rate of extinction increased dramatically as forests and plains were converted to farms and cities. The rate has continued to accelerate—since 1900 technological man has caused the extinction of about 100 species of mammals, birds, reptiles, amphibians, and fishes, and we are not yet through the century. Whereas one species disappeared every 1,000 years before man, a vertebrate species is lost about every ten months today (Myers, 1979). Even this rate is misleading, because dozens of invertebrates and plant species are lost for every vertebrate that disappears. Taken altogether, it is estimated that 1,000 species of plants and animals became extinct each year during the

late 1970s; about one every nine hours (Myers, 1979). And even this rate is accelerating (Lovejoy, 1980a; Myers, 1981).

Extinctions are occurring worldwide, but they are most frequent in the moist and wet tropics. One-quarter to one-half of the world's species are found in these habitats, but they occupy no more than 6% of the earth's land area (CEQ, 1980). The species-rich tropical rain forests of Latin America, Southeast Asia, and Central Africa are being destroyed to feed the lumber industries of the developed nations and the firewood requirements of less developed nations, to supply land for pastoralists and farmers, and to provide foreign exchange to local governments (Lovejoy, 1980b; Myers, 1979). Many of the wild species that live in those forests are not yet known to science. Since the beginning of recorded history, man has discovered and described approximately 1.5 million species of plants and animals. Yet the rate at which new species are still being discovered, the large areas of the globe that are still relatively unstudied, and the known diversity of species in some habitats suggest that between 8 million and 10 million species are alive today. This means five out of six living species have not yet been discovered. Many are being lost even before they are known. Their extinction is doubly tragic because man has not had the opportunity to assess their potential value (Raven, 1981). It is estimated that more than 25,000 species of plants and over 1,000 species of animals are in danger of extinction (IUCN, 1975; Lucas and Synge, 1978; Thornback and Jenkins, 1982).

In these days of budget cutting and economic reorganization, particularly in Washington, D.C., it is often claimed that the benefits derived from various governmental programs must outweigh the dollars that would be saved by cutting them. For example, a particular program might yield five dollars paid as taxes for every dollar spent by government. A five to one return is considered an extremely good investment. However, no government program has yielded direct and indirect benefits as significant as those provided by wild plants and animals; these benefits are seldom acknowledged and always undervalued in political decisions.

Foods

One of the benefits provided by wild species is food. People in every nation consume wild species. For example, over 72 million tons of fish and shellfish were caught worldwide in 1976. Marine species accounted for 61 million tons of the total, 9.6 million tons were freshwater species, and 1.4 million tons were diadromous species that live in both fresh and salt water. In developed countries such as the United States, wild species in the forms of fish and shellfish account for 12% of the animal protein consumed and 7% of our total protein intake.

In many nations, particularly those in the tropics, wild birds, mammals, reptiles, amphibians, fruit, and vegetables are also principal sources of food (IUCN, 1980). Unfortunately, the importance of wild species is seldom documented in the publications of nutritionists and government statisticians concerned with food. Food sources such as squirrels, rabbits, raccoons, iguanas, bandicoot rats, beetle grubs, bee larvae, wild rice, berries, wild mushrooms, and roots simply do not enter the picture when nutritionists are discussing human diets (IUCN, 1980). If this were not the case, perhaps governments would be more concerned about protecting these wild species.

On a per capita basis some forest cultures in tropical countries have for generations consumed as much or more protein from wildlife than citizens of some industrialized nations do from livestock, yet in the rush for development the productive forest habitat is being cleared to make room for ranches and farms. In much of Africa between 50% and 70% of the animal protein consumed by humans is bush meat—wild animals caught, snared, trapped, or shot to put meat on the table (Coe, 1980). One of the animals most frequently eaten in West Africa is a large rodent called a canecutter, *Thryonomys swinderianus* (Asibey, 1969, 1974). The number of canecutters eaten each year runs into the hundreds of thousands. Probably over 500,000 kilograms of protein in the form of canecutters are put on the table, but few government programs consider the conservation of these rats. The same is true of the much-hunted agouti and pacasat in America, or the bandicoot rat and deer mouse of southern Asia.

Iguanas are another important source of protein in Central American nations from Mexico to Panama, but little recognition has been given to the importance of these lizards in human diet and little has been done to conserve them. Even developed nations consume quantities of wild animals. For instance, over 3 million kilograms of frog legs were sold in the United States in 1973 (King, 1978b).

Wild plants also provide many local foods. Some—such as wild rice, truffles, durian (*Durio zibethinus*), persimmon (*Diospyros virginiana*), wild strawberry tree (*Arbutus unedo*), and genip (*Genipa americana*)—are much sought after by local people. Others, such as wild tamarind and wild coconut, are exploited for an international market.

Wild animals and plants also represent the genetic material from which crops and livestock are bred (CEQ and Department of State, 1980; Myers, 1979). Selective breeding of the red jungle fowl of southern Asia gave rise to the chicken. The agrimi goat of the Middle East is the ancestor of the domestic goat. The greylag goose is the progenitor of the barnyard goose. All livestock and all domestic animals were bred from wild species. None could have been developed without access to their wild ancestors. The same is true of crops and ornamental plants.

Many wild plants have given rise to multiple cultivated varieties that are commercially valuable. A single weedy plant from western Europe, *Brassica oleracea*, was the stock from which seven vegetable crops were bred: head cabbage, collards, kale, broccoli, cauliflower, brussels sprouts, and kohlrabi (Mathias, 1978). This is not an isolated phenomenon. There are over 2,000 named varieties of Japanese camellias and an equal number of roses and tulips.

Only a few wild animals have been domesticated, and an equally sparse list of wild plants gave rise to the crops on which a majority of the world's people depend (CEQ, 1980; Ehrlich and Ehrlich, 1981). Not many birds are habitually found among man's livestock—chicken, turkey, geese, one or two species of duck, guinea fowl, peafowl, and pigeon. Other pandemic (widespread) species of livestock are even fewer in number—cattle, horses, goats, sheep, donkeys, pigs, and rabbits. Similarly,

FIG. 4.1. A Nepalese woman milking a yak, a livestock species virtually the same as its wild relatives. Photograph by M.J. Odell.

how many grain crops are regular features of man's diet—wheat, corn, rice, sorghum, oats, barley, millet, rye? More than half of the world's food-producing agricultural lands are involved in supplying the first four of these crops alone (Myers, 1979).

If being dependent on only a few species for our major crops were not bad enough, many of those crops were developed from only one or two varieties, or even one or two specimens (Myers, 1981). For example, only six plants from a single site in Asia were used to launch the North American soybean industry, and 72% of the United States' potato crop is derived from four varieties (IUCN, 1980). All cultivars of such limited genetic diversity are susceptible to pests and disease (CEQ, 1980; Ehrlich and Ehrlich, 1981; Myers, 1979).

In 1845 and 1846 potato blight destroyed the agricultural economy of Ireland, resulted in the deaths of an estimated 2 million people, and forced the emigration of thousands of Irish families (CEQ, 1980). It took eighty years to find a wild potato, in Mexico, that was resistant to the blight and to breed that resistance into the North American and European crops. Similarly, in 1970 five inbred lines

of corn produced 70% of the seed corn used in the United States (Myers, 1979). That year southern corn leaf blight swept the United States and reduced corn yield by about 15% (CEQ, 1980). Only by having a variety of corn genetic stocks from which to breed resistant strains was it possible to guarantee better yields in subsequent years. Another blight, lethal yellowing, is creeping across the Caribbean and destroying coconut trees, which yield plant protein, sugar, coconut milk, and oil for human consumption, as well as fiber, thatch, and charcoal. In 1978 an estimated 375,000 coconut palms had died of the disease in Florida, and 1,000 a day were dying in Jamaica. In 1982 the blight reached Yucatán and threatened to spread throughout Latin America, where half of the coconut crop is used for subsistence. Fortunately, varieties from Malaysia have thus far proved resistant to the yellowing disease, so it may be possible to replace the lost trees if the blight cannot be stopped by other means.

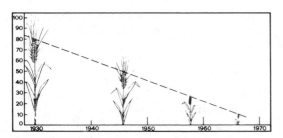

FIG. 4.2. *The decline in percentage of primitive cultivars in the Greek wheat crop represented here is typical of most crops in most countries. Graph from IUCN, 1980.*

Corn, potatoes, rice, and wheat feed more people than the next twenty-six major crops combined (Myers, 1979). A pandemic, virulent blight attack on any of these four crops would have devastating effects and cause the deaths of untold millions of people and the economic ruin of others. In the United States alone, corn is a $20-billion-a-year crop (Raven, 1981). It takes ten to twelve years to develop a new disease-resistant variety when suitable breeding stock can be found, and that new resistance lasts only five to fifteen years before it succumbs to a more virulent strain of disease (IUCN, 1980; Myers, 1979, 1981). This constant battle to stay ahead of evolving pests and disease is dependent on the conservation of wild and semiwild varieties of rice, wheat, corn, sugarcane, tea, soybean, bananas, coconuts, cacao, and all other valuable crops. These wild relatives of our crop—and livestock—species are invaluable as a reservoir from which to breed new resistant varieties, or ones with improved flavor, improved nutrition, greater yields, different growing seasons, or adaptability to different soils and climates. One disease-resistant wild wheat from the Middle East boosted wheat production in the United States by $500 million a year (Myers, 1981).

In the late 1970s a weedy relative of domestic corn was discovered in Jalisco, Mexico (CEQ, 1980; Raven, 1981). Unlike domestic corn—which is an annual crop—this relative, *Zea diploperennis*, is an ever-growing perennial. Hybrids between the two might lead to the development of corn orchards, where the same corn plants would grow year after year without dying after harvest. The savings in energy from not having to plow and seed the crop

clearly make this goal worth pursuing. Hybrids also may be more resistant to the many diseases that presently infect the annual corn crop (Myers, 1981).

Many such wild relatives of the major food crops are being lost as modern agriculture actively discourages the use of genetically diverse local food crops in favor of high-yield monocultures. While such strategies have greatly increased the yield of foodstuffs, they have also reached the point that continued replacement of the local cultivars threatens their increased productivity (CEQ, 1980; Ehrlich and Ehrlich, 1981; Myers, 1979). High-yield monocultures depend on intensive use of pesticides and fertilizers. And they are susceptible to many weed and insect pests and to diseases that can only be overcome by crossbreeding with the resistant local cultivars.

In the Andes of Peru and Bolivia dozens of species and varieties of potatoes are found. They yield tubers of many shapes, sizes, colors, textures, flavors, cooking properties, and harvest seasons. Yet modern agriculture threatens to replace many of these local potato crops with one or two varieties. The genetic diversity represented by these local crops will be lost unless Andean farmers are encouraged to keep them in cultivation alongside the new varieties. Similarly, wild native relatives of coffee, okra, and pea are being lost in North Africa. Wild cucumber, eggplant, radish, squash, black-eyed pea, and yam are disappearing in Asia. Wild asparagus, cabbage, beet, lettuce, wheat, and turnip are vanishing from the Mediterranean region. Wild varieties of bean, lima bean, pepper, corn, potato, sweet potato, pumpkin, and tomato are being lost in Latin America.

These observations should not be construed as opposition to high-yield agriculture, because they are not. But they are a caution against allowing the rush to intensive agricultural methods to carry away the very cultivar diversity on which its high yields depend. In many instances the appropriate use of high-yield cultivars and intensive agricultural methods will alleviate the need to convert more wild habitat to farm use.

Although only a relatively few species of plants and animals produce a majority of the agricultural foodstuffs, many others are used locally and could become significant sources of human food if properly developed (CEQ, 1980). One such species is the winged bean, *Psophocarpus tetragonolobus*, of New Guinea and Southeast Asia (National Academy of Sciences, 1975). It produces a large, high-quality protein bean, and its leaves, shoots, flowers, and pods can be eaten as greens. It also produces an underground tuber with a taste like early-season potatoes; but unlike potatoes, which have a protein content of 3% to 7%, the winged bean tuber is 20% protein. Many other local crops could be developed for wide consumption by both people and livestock.

Still other species that are not dependent on fresh water could be used for food where fresh water is limited. Virtually all our grain crops and the majority of our fruit crops require fresh water in the forms of rain, groundwater, or artificial irrigation. However, the Seri Indians of northwestern Mexico used to harvest the abundant seeds of a submerged grass, *Zostera* sp., that abounded in the brackish estuaries where the Colorado River flows into the Gulf of California (Felger and Moser, 1973). The seeds were ground into flour for tortillas and bread. Unfortunately, the United States has diverted so much water away from the Colorado River that the *Zostera*, although still there, is no longer abundant enough to be easily collected in quantity.

Another salt-tolerant plant, Palmer's grass, *Distichlis palmeri*, found in the northern Gulf of California, produces a grain used by the Cocopa Indians (Neary, 1981). An inedible, salt-tolerant wild tomato, *Lycopersicon cheesmanii*, that is found in the Galapagos Islands has been crossed with the commercial tomato, *L. esculentum*, which cannot tolerate salt, to produce tasty fruit that can be grown in 70% sea water (CEQ and Department of State, 1980). These and other salt-tolerant species could be used to produce alternate crops in coastal areas and in deserts, where salt water is available and fresh water limited.

Of the approximately 300,000 species of higher plants, only a few dozen have become major food crops (CEQ, 1980). Most of these were first brought into cultivation by Neolithic peoples, well over 5,000 years ago. The future well-being of man depends on conserving every plant that has the po-

tential of becoming a new important crop or of improving an existing crop (IUCN, 1980).

Medicines

Thousands of wild species of plants and animals are used in the production of medicines (CEQ, 1980; Ehrlich and Ehrlich, 1981; Myers, 1979). The storehouse of chemical compounds found in wild species is an irreplaceable source of medicines and other drugs—curare, cocaine, penicillin and quinine are but a few. The development of many synthetic drugs—aspirin, cortisone, and oral contraceptives among them—was dependent on chemicals or their precursors derived from wild plants and animals. Aspirin was first isolated from willow bark in Europe. It is highly doubtful that many manufactured pharmaceuticals would be in use today had their natural prototypes not first been discovered in wild plants and animals.

Only about 5,000 flowering plant species (2% of the total) have been tested for alkaloids of medical value, yet they have yielded a surprising array of pharmaceuticals. Some 35,000 higher plant species have been screened for anticancer drugs. Each year approximately 1,500 new compounds are reported from plants, and about 300 of these have biological activity and potential use in medicine. About 1,000 antibiotics have been produced from plants (mostly fungi) and bacteria (Ehrlich and Ehrlich, 1981). Over 40% of all prescriptions written in the United States each year contain a drug of natural origin as the principal active ingredient (Farnsworth and Morris, 1976). Of seventy-six major pharmaceutical compounds obtained from higher plants, only seven can be produced synthetically at competitive prices. For example, reserpine, used in the treatment of high blood pressure and nervous tension, can be produced from *Rauwolfia* plants at $0.75 per gram, but synthetically for $1.25 (Myers, 1979).

In 1972 the United States imported $24.4 million worth of plant products for pharmaceutical and cosmetic uses (Myers, 1981). Similar imports by all developed nations totaled $74 million. In the United States alone the value of medicines produced from higher plants is estimated to be about

$3 billion annually; the value of all medicines of natural origin is $10 billion a year—and it is rising each year. It is impossible to predict which species might be the source of the next important medicine.

The Madagascar periwinkle, *Catharanthus roseus*—a flowering ornamental that is grown widely in warm parts of the United States, Europe, Australia, Asia, and Latin America—produces two compounds, vincristine and vincablastine, used in treating some types of cancer (Ehrlich and Ehrlich, 1981; King, 1978a; Mathias, 1978; Myers, 1979). Vincristine, had worldwide sales in 1976 of $22 million. Used in combination with other chemicals, it achieves 80% remission of Hodgkin's disease, 99% remission of acute lymphatic leukemia, and 50% to 80% remission in several other types of cancer. This flower also produces raunasine, a drug used to combat hypertension.

The seed of evening primroses, *Oenothera* sp., is an important source of gamma-linolenic acid, which is an essential fatty acid and a precursor of prostaglandin production (Raven, 1981). Essential fatty acids must be available in foods; they cannot be synthesized in the body. They are important in the formation and functioning of cell membranes. Prostaglandins are hormones that control the functioning of body organs. A deficiency of essential fatty acids leads to diseases of the circulatory system, arthritis, and nervous disorders. Gamma-linolenic acid is the most active of the essential fatty acids used in treating such a deficiency. The only other ready source for this compound is human milk. But approximately 8% of the species of evening primroses are threatened or endangered (Raven, 1981).

Researchers in England have just discovered a hawkmoth that is the source of an anticancer compound (Miriam Rothschild, personal communication). Ara-C, a compound from a Caribbean sponge, *Tethya crypta*, is an effective inhibitor of leukemia and some other cancers (King, 1978a). Horseshoe crabs now supply enzymes that are used in blood tests for some human hereditary diseases. Other compounds from sponges, coral, sea anemones, marine worms, sea cucumbers, and starfish might be useful in combating cancers, hyperten-

FIG. 4.3. *The Madagascar periwinkle is the source of two important anticancer compounds and an antihypertensive drug. Photograph by S. Shumway.*

sion, and cardiovascular disease. A number of these animals are already valuable sources of antibiotics (Ruggieri, 1976).

In the last ten years Oriental poppies, *Papaver orientale*, have been discovered to produce a chemical precursor to the important analgesic codeine (Raven, 1981). The precursor cannot readily be used to produce the undesirable and much abused heroin, and it causes convulsions. Substitution of this alternate species for opium poppies, *P. somniferum*, in opium-growing nations will guarantee farmers their earnings while helping to alleviate one of the world's major drug abuse problems.

The painted daisy, *Chrysanthemum coccineum*, of Southwest Asia is the source of pyrethrum, one of the most widely used biodegradable pesticides, which is relatively nontoxic to birds and mammals. Another natural repellent or pesticide is rotenone, extracted from the roots of several genera of tropical leguminous vines and shrubs (Myers, 1979). The only way we can be sure that these and other species will be available to benefit future generations is to conserve them now.

Industrial Products and Benefits

Wild species also have many industrial uses. For example, forests produce timber, fiber, paper, fabrics, resins, gums, dyes, waxes, oils, aromatic compounds, mulches, and a variety of other materials that annually earn more than $100 million for each of thirty different nations, including the United States (IUCN, 1980). Southeast Asia produces over 20% of the world's hardwood logs, most of which are processed into plywood, including over 90% of the plywood imported into the United States (Mathias, 1978).

One of the most important forest products, often overlooked in developed countries, is fuel. Many regions of the world are being deforested to supply firewood. It is estimated that 80% of the people of Thailand cook over wood fires. Annual consumption of firewood in Thailand is 1.34 cubic meters per person (Suvanakorn, 1980). Total consumption by the 45 million Thai is 7 million cubic meters. Of this total, only 3 million cubic meters is produced in Thailand; the rest is imported (Suvanakorn, 1980). One result of this shortage is rampant wood poaching, which even invades national parks. Similar situations exist in many tropical nations. In North Africa some tribal women spend in excess of 300 days a year actively searching for wood with which to cook (IUCN, 1980). It is, however, often possible to conserve native forests by planting fast-growing wild species from other areas. Some of these alternate tree species are legumes capable of producing 50 to 100 cubic meters of wood per hectare each year (National Academy of Sciences, 1980; Vietmeyer, 1978).

Oil palms, Elaeis guineensis, native to Africa have been cultivated extensively in the tropics of Africa, Asia, and Latin America to produce oils for soap, candles, and lubricants and also edible oils for margarines, confections, and pharmaceuticals. Although such uses are important, these oils will become increasingly significant as a fuel as world petroleum reserves become depleted. Oil extracted from the dried meat of the coconut, Cocos nucifera, has a high palmitic acid content demanded in the manufacture of soaps and detergents.

One new alternate crop is the jojoba bean, Simmondsia chinensis, a native of the deserts of southwestern North America (CEQ, 1980; Ehrlich and Ehrlich, 1981; Myers, 1979). It is now being cultivated in California, Arizona, Israel, and Saudi Arabia for the liquid wax in its seed. This wax is essen-

tial for high-temperature lubrication of automatic transmissions and high-temperature cutting of metals. Until jojoba came into cultivation, similar waxes were obtained from sperm whales. The California meadow foam, Limnanthes sp., is yet another wild species with seeds that produce sperm whale–like liquid wax (Mathias, 1978). And the creosote bush, Larrea tridentata, so familiar from western movies, is the source of a resin that prevents butter and margarine from turning rancid (King, 1978a).

Other alternate sources of hydrocarbon chemicals similar to those found in petroleum are some of the latex-producing trees and shrubs of the families Euphorbiaceae, Compositae, and Sapotaceae (Ehrlich and Ehrlich, 1981; Myers, 1979). Unlike petroleum, these latex compounds are free of sulfur and other contaminants. The Amazonian rubber tree, Hevea brasiliensis, is the source of most natural rubber, but guayule, Parthenium argentatum, a little-known desert shrub of the southwestern United States and Mexico, is another source (National Academy of Sciences, 1977). During the 1940s the United States produced 15 tons of guayule rubber daily from plantations and two factories in California. Today Mexico commercially produces guayule rubber in limited quantities, but plans soon to produce 30,000 tons a year. Although rubber produced from petroleum has surpassed natural rubber for many uses, Hevea and guayule rubber remains in great demand. One-third of the world's total production of rubber comes from Hevea, but by the late 1980s and 1990s the demand for natural rubber will exceed supply. Twelve species of Euphorbia native to Brazil have been identified as possible sources of phytochemicals that can be used to produce gasoline and other fuels (Myers, 1979).

Animals that inhabit forests and other natural habitats also produce valuable products for industry (King, 1978b). Beavers, mink, raccoons, otters, bobcats, lynx, ocelots, jungle cats, weasels, musk deer, civets, ostriches, rheas, crocodiles, pythons, boa constrictors, iguanas, monitor lizards, morpho butterflies, lac insects, and many other species supply furs, skins, feathers, felt, iridescent wings, oils, essences, and lacquer. By 1975 imports were over $1 billion (King, 1978b).

Other species of wild animals supply chemicals useful to industry. The Japanese insecticide Padan is derived from nereistoxin, a natural poison found in a marine polychaete worm (King, 1978a). Padan is effective against strains of insects that have become resistant to organophosphate and organochlorine compounds. Pests such as Colorado beetle, Mexican bean beetle, cotton boll weevil, rice stem borer, cabbage butterfly, and diamondback moth are all treated with Padan.

Most of the species that are presently important commercially are abundant. But if they are overexploited, they may become depleted or endangered. Occasionally, critically endangered species become the bases of industries, are artificially propagated, and as a result are saved from extinction. One such example is the golden hamster, *Mesocricetus auratus*, of Syria (Walker, 1975). Only two specimens have ever been found in the wild; the first in 1839, and the second in 1930. The second was a female that gave birth to twelve young, and it is from this litter that all the golden hamsters of the pet trade have descended. Sales of golden hamsters worldwide, including their food and cages, exceed several hundred thousand dollars annually. Another example is the Franklin tree, *Franklinia alatamaha*. It was originally discovered growing on the banks of the Altamaha River in eastern Georgia, United States, in 1765. Seeds and plants were collected in 1773 and 1778, but the tree was never seen in the wild after 1790 (Little, 1980). Today it survives as an ornamental tree cultivated for its beautiful 7.5-centimeter-wide blossoms, which appear each fall. Sales of *Franklinia* probably do not exceed several thousand dollars annually, but they are sufficient to guarantee the survival of the species. The commercial meristem cloning of rare orchids and succulents satisfies a multimillion-dollar-a-year industry and promises to save many of these species from extinction as well, though it must be remembered that such propagation methods conserve the genetic diversity of only one or a few individuals.

Ecological Services

Wild species also provide important ecological benefits (Ehrlich and Ehrlich, 1981; IUCN, 1980; Myers, 1979). Although a species may appear to be useless, many species and the ecosystems they occupy are critical to the well-being of economically important species. For instance, mangroves and the swamps they form are considered by many to be little more than breeding grounds for mosquitoes and other biting insects. Many mangroves were destroyed as coastal areas were developed. The trees were cut for timber or firewood, or simply to get rid of them. Bulkheads were erected and filled with earth where the mangroves once stood. A decline in the productivity of local fisheries often accompanies the loss of mangroves, for their fringing swamps are the nurseries of many commercially important fish and prawn species (IUCN, 1980). Similarly, coastal estuaries and shallow embayments in the temperate zones sustain fisheries resource species such as cod, herring, plaice, and sole, which are a direct benefit to mankind.

Because man is an active manipulator of his environment, he often thinks of human-modified habitats as "improved" and untouched natural habitats as "wasteland." As a consequence, the many ecological services provided by wild species and natural areas are often overlooked; but they are extremely expensive to replace, if they can be replaced at all. Clearing watershed vegetation is a frequent and costly example of manipulative destruction of wild species and habitats.

Watershed forests, shrubs, swamps, and marshes have been lost all over the globe as land has been cleared for agriculture, cities, firewood, and a variety of other reasons (IUCN, 1980; Myers, 1979). Watershed vegetation performs a number of essential functions. Terrestrial plants and leaf litter hold water, slowly releasing it to the streams. Without this vegetation, rain runs off the land so fast that erosion often results. Fertile soils are washed away. Farms become less productive. Rivers become laden with sand and silt, which settles out as shallow sandbars in the slower currents downstream or fills reservoirs of calm water behind dams and barrages. In addition, the speed of the runoff may cause rivers to flood higher and more frequently than before the vegetation was lost. Some of these problems can be controlled, but only by expensive engineering efforts. Contour terracing of the land and construction of dikes, levees, and dams will slow soil erosion from denuded slopes and prevent downstream flooding. Frequent dredging of the river will keep sandbars out of the main channels.

But all these remedies are expensive in more than just dollars. Each causes additional problems.

The fast-flowing rivers of North America once contained 60% of the world's unionid clam species, but because siltation occurred behind dams constructed to control flooding and improve navigation, an estimated 80% to 90% of these species are now extinct or endangered (David Stansberry, personal communication).

Watershed vegetation, both terrestrial and aquatic, also extracts large quantities of nutrients from the water before releasing it downstream. This helps slow the eutrophication of the stream or river. Without the vegetation, and especially where stream channels have been straightened and dredged deeper, the nutrients often are flushed downstream to pollute and cause stagnation of the water. This problem cannot easily be controlled by technological means.

The least expensive method of correcting flooding, siltation, and nutrient-loading problems is to revegetate the watershed. This is possible, however, only if suitable species—preferably the original species—are still available (Vietmeyer, 1978), and only if urbanization was not the reason for denuding the watershed in the first place—relocation of cities usually is neither economically nor culturally feasible. It is far less expensive to prevent the loss of plant cover in the first place. This can be done by recognizing the services a functioning watershed ecosystem—or other natural ecosystem—performs, while it still exists.

A variety of other benefits are also provided by communities of wild species. Amelioration of climate is one. Removal of forest cover in many parts of the world has increased insolation, reduced availability of water, and produced deserts (CEQ, 1980; IUCN, 1980).

Free essential services are not always performed by whole biotic communities. They are also delivered by single species or small groups of species. For example, legumes, together with the *Rhizobium* bacteria living symbiotically in their roots, extract nitrogen from the air to synthesize compounds that enrich the soil for other plants (Myers, 1979; Raven, 1981). Without nitrogen, plants cannot produce protein. Some of the approximately 18,000 species of legumes are economically impor-

tant crops (for example, alfalfa or lucerne, beans, peas, and soybeans), others can become weeds (white clover and kudzu vine), but all play this important role in combination with *Rhizobium*. Very few other plants or free-living bacteria can fix nitrogen. Can man afford the loss of free fertile soil that might result from the extinction of even one legume?

Other wild species are important as pollinators of crops and ornamental plants, and as predators and parasites of crop and forest pests (Ehrlich and Ehrlich, 1981). An estimated ninety crops in North America are dependent on insects to pollinate them. Without the various honeybees, bumblebees, solitary bees, wasps, hover flies, soldier flies, and other insects that carry out the pollination, their flowers would not be fertilized, reproduction would not be complete, and there would be no yield of fruit or vegetables. An additional nine North American crops have increased yields because of insect pollination.

Many insects prey on crop or forest pests. Ichneumonid wasps, praying mantises, ladybugs, and lacewings are but a few of these predators valued by North American gardeners. Some pests that are not easily controlled by chemical means might well be reduced to negligible levels by using other species for biological control. For instance, a parasitic wasp from India, *Pediobius foveolatus*, is being studied as a possible control for the Mexican bean beetle, a costly and persistent pest of legume crops in the United States (CEQ, 1980). Insect predators in tropical forests are so numerous that farms carved out of these habitats often are better off relying on these living pest control agents than using chemical pesticides (Ehrlich and Ehrlich, 1981). The chemicals kill both predators and pests; but if the pests develop a tolerance to the chemicals, they may become even more numerous than they were in the beginning.

Scientific Value

Many wild species have great scientific value (Myers, 1979). The 8 to 10 million species alive today and the ecosystems they are part of are living laboratories for science. Many significant advances in science have come from studying seemingly unimportant species. For example, an understanding of human genetics came about through research on the genetics of fruit flies and horseshoe crabs. A great deal of what we know about plant genetics resulted from the study of sweet peas. Developmental and reproductive biology began with the study of sea urchin eggs, frog eggs, and salamander larvae. Research on squid and electric eels gave insight into the workings of nerves. The development of sonar stems from research on the echo location of bats.

It is virtually impossible to predict which species will provide another scientific advancement (CEQ, 1980, 1981). Studies on the hibernation of black bears led to the development of an improved low-protein, low-fluid diet for human patients with kidney failure (King, 1978a). Research on the octopus is an aid to understanding aging. A microscopic planktonic organism, *Umbilicospirea* sp., has been discovered to store uranium (in the sugars it produces) in concentrations 10,000 times greater than its natural occurrence in the sea; study of this organism might lead to a method of cleaning up radioactive spills or of producing some nuclear fuels (King, 1978a; Myers, 1979). An endangered desert pupfish shows great tolerance to extremes of temperature and salinity and is being investigated as a model for research on human kidney diseases. Nine-banded armadillos are being used in studies on leprosy (Myers, 1979). American alligators are subjects for investigations of antibody formation. Research on elephants is elucidating some aspects of atherosclerosis. Tissues from pigs and sharks have led to development of new skin graft techniques. The examples are endless.

Cultural Values

Living natural resources provide many cultural, emotional, aesthetic, and recreational benefits. Wild species offer delight, inspiration, and instruction to man. Species such as tigers, butterflies, robins, roses, whales, and lotuses are cultural resources. What a loss it would have been to human culture if they had become extinct before storytellers, artists, poets, and songwriters had a chance to observe and incorporate the images of these organisms in their works. The extinction of any species carries that same potential.

Threats to
Species Survival

The single greatest threat to the survival of wild species is habitat destruction (IUCN, 1980; Myers, 1979). Habitats are being lost at an ever-increasing rate as forests are felled, prairies plowed, and wetlands drained—all to create products for industry or to make room for more farms and cities. Pollution, the poisoning of the environment, is another form of habitat destruction. No species is born free. They are all captives of their environment. They live only where their needs are met. The loss of a mature rain forest, with its hole-filled old trees, spells the death of the hornbills that need the holes for nests. Loss of extensive mature hardwood forests caused the extinction of the ivory-billed woodpecker, *Campephilus principalis*, in the eastern United States. Loss of open grassland contributed to the extinction of the heath-hen, *Tympanuchus cupido cupido*, in the northeastern United States; and reduction of open prairie has endangered Attwater's prairie chicken, *Tympanuchus cupido attwateri*, in Louisiana and Texas (Greenway, 1967).

The second most frequent cause of wild species extinction is commercial overexploitation (IUCN, 1980). Uncontrolled exploitation can lead to the extinction of abundant species in as little as thirty to forty years. The passenger pigeon, which originally numbered in the billions, disappeared after about forty years of intense commercial exploitation. The same is true of a number of other wild animals and plants (King, 1978b).

It should be noted here that sport hunting usually is not a threat to the survival of wild game species. It generally is sufficiently well regulated to prevent overexploitation of the game. Occasionally, however, habitat management that benefits game species causes the decline or extinction of nongame species. Modification of habitat to benefit waterfowl, for instance, was one factor contributing to the decline and endangerment of the dusky seaside sparrow, *Ammospiza maritima nigrescens*, in the southeastern United States. Subsistence hunting to put protein on the table often is not controlled and can cause the extinction of species, particularly when coupled with a growing human population and a dwindling plant or animal resource.

A third important factor contributing to the extinction of wild species is competition from introduced exotic species (CEQ, 1980; IUCN, 1980). When a species is released into a habitat or ecosystem to which it is not native, it may outcompete the native species for some limited resource that is essential to their survival (Courtenay, 1978). Competition from introduced exotics is the most frequent cause of extinction of oceanic island species. Exotics such as goats, rabbits, mongooses, cats, dogs, pigs, donkeys, macaques, rats, trout, largemouth black bass, and *Talapia* have caused the extinction of hundreds of species and the endangerment of many others (Allen, 1942; Greenway, 1967; Harper, 1945).

There is one additional major cause of extinction—the extinction of coevolved or mutualistic species. The loss of one species frequently has a falling-domino effect that leads to the extinction of other dependent species. Parasite species will become extinct if their obligate host organisms disappear. The loss of a species that provides the only food, habitat, or ecological requirement of another species will bring about the demise of the second species as well. The Mauritius dodo provides a good example of such dependence (Temple, 1977). The dodo fed on the fruit of the *Calavera* tree, whose seed is enclosed in a kernel so tough that it will not germinate unless it is first eroded in the digestive tract of the dodo. No seeds have sprouted on Mauritius since the dodo became extinct in 1680. The few trees that survive are 300 years old and stand as grim reminders of the passing of the dodo. Similarly, the extinction of any large vertebrate will probably result in the loss of several species of invertebrates—the extinction of most large mammal species will cause the extinction of a specific dung beetle (Robert E. Woodruff, personal communication). The disappearance of any rain forest plant species probably results in the extinction of ten to fifteen other species, primarily animals and microorganisms (Raven, 1981).

Other causes of extinction exist (for example, disease and natural catastrophes such as volcanic eruptions, hurricanes, or droughts). They are minor, however, compared with habitat destruction, commercial overexploitation, introduction of exotic species, and the loss of coevolved mutualistic species.

Conclusion

Human-caused extinction of wild species is probably the most important problem facing mankind today. Growth of the human population, coupled with the expansion of agriculture and the spread of cities, has accelerated the rate of extinction, and that rate is continuing to climb. All humans are dependent on wild species and the crops and domestic animals derived from them for food, medicines, fibers, shelter, fuels, industrial products, ecological services, ecosystem stability, scientific benefits, and cultural, aesthetic, and recreational values.

An extinct species represents lost opportunities for economic, scientific, and cultural benefits. The dodos became extinct largely from being killed for food by European sailors. If an effort had been made at that time to conserve the dodo, to breed it selectively, we might have Kentucky-fried dodo franchises in every town. A domestic offspring of the dodo might have become a better barnyard fowl than the chicken derived from the red jungle fowl. Unfortunately, the opportunity has been lost forever. Man cannot afford such losses. What future famine might occur because a disease-resistant primitive cultivar or wild species is no longer available for crossbreeding? What future disease might go uncured or untreated because some pharmaceutical species has disappeared? How much productive land will be converted to desert because the plants and animals that formerly lived on it were stripped away?

The present and future well-being of man depends on his conservation of wild species. Development cannot be sustained in the absence of conservation. The benefits of conserving wild species outweigh the costs, and conservation costs are far less than the cost of not conserving.

Discussion

QUESTION. Endangered species programs often seem to exist in isolation. Can they be made more useful to developers and decision makers? Can they be made more open to public participation?

ANSWER. First, let me address the issue of the developers and decision makers. Endangered species programs often produce inventory maps spotted with high-priority areas for rare species protection. Decision makers need maps that are the exact opposites of these. Such maps should show the least important areas, the places that can be "chewed up" as much as necessary. If you want to protect an area, the best way is often to point to a good alternative spot for development. Many such inventories are done in states where public monies are not available for them. The Audubon Society has successfully raised money for surveys from constituencies as diverse as power companies, traditional garden clubs, and conservation organizations. Industries have been active participants in such studies because they want to know where they can develop. Inventory maps become conflict-avoidance tools for them. I think we have captured their interest and focused it in a very productive way.

Now the public. What have we seen over the last four centuries in the United States? First, the early eastern colonists farmed and grazed, removing the woodlands everywhere except the rocky slopes and mountains. Then the west was won and New England abandoned agriculture. The land naturally reforested, but the new forests were impoverished compared with the old. The same cycle is about to be repeated. When the tropical forests are gone, the price of timber is going to become prohibitive. Industry will start cutting that eastern forest again. New England is on the brink of massive change. Ecologists need a record of what it is like now, because without that record there will be no objective base for land management decisions. All we have now are inaccurate, qualitative descriptions. If we had a complete atlas of the flora, we could see how change takes place. We could monitor and control it.

What do we need for such an essential project? First we need a few committed plant scientists. Plant scientists who know their flora and are willing to interact without competition. But then we need the public. We must get schoolteachers, kids, everybody, to take a part in unraveling the mystery of the landscape. Everyone can feed information to the central core of scientists for clarification. Such a project awakens interest and public awareness at the same time that it provides an extremely valuable service to both basic science and environmental management.

QUESTION. It seems more economically worthwhile to invest in improving the crop varieties we have now than to hope that some wild species will have economic value. I think there are other good reasons for species preservation, but I am uncomfortable with this one. Can we really claim economic benefits for agriculture from wild species?

ANSWER. It is hard to be sure what an "improvement" is. The economics of agriculture change, particularly the relative costs of muscle and technology. This is a very important truth. It means that we will need different kinds of crops in the future, ones we can only guess at now. The plantations of the Malaysian Peninsula export more rubber than any others in the world. In 1860 they were growing coffee. Rubber did not become economically viable until the turn of the century. The demand for new crops and the basic economics of farming are changing radically. If we do not have entirely different species and varieties to exploit, we will be in extraordinarily serious trouble.

Today there is a tension between simplification and maintenance of diversity in all kinds of crops, animals as well as plants. On the one hand, to maximize present productivity we need uniformity. Let's use the rubber tree as an example again. All commercial rubber trees are the product of only three genotypes, a root stock genotype grafted to a stem genotype grafted to a crown genotype. All commercial rubber trees are genetically the same— and they are all very productive. You spray on the insecticide, lash on the fertilizer, and you get something like fifteen times the natural level of production. But now the economics are changing because the costs of the pesticides and fertilizers are rising out of sight. We need very different rubber trees. The Rubber Research Institute in Malaysia is sending expeditions back to the Amazon to get new genetic material because the whole rubber industry, which is worth billions of dollars every year, is based on just those three genotypes from a single seed collection. The institute realizes how badly it

53

needs new strains, preadapted in some way to these changed economic conditions.

QUESTION. How do you approach people whose job it is to exploit resources in the present? Can you get them to think about long-term goals?

ANSWER. You have asked about a very difficult problem. I start toward an answer by trying to convince people that we are not separate from the ecosystems in which we live. We are an integral part of them. Conservationists have to persuade people to look to the future and to cooperate in the present. The era of freewheeling entrepreneurs in land or energy development is rapidly coming to a close. For instance, when the rain forest goes the humus in the soil will be oxidized and will add significantly to the carbon dioxide in the atmosphere. This will be on top of the carbon dioxide already being released as we burn fossil fuels. Together, these will greatly increase the greenhouse effect, raising the seas 6 to 8 meters. Most U.S. cities are on the coasts. What will happen to Boston, New York, Washington, Philadelphia, New Orleans? They will be flooded. We cannot afford to build new cities 20 meters higher on the slopes. So all coastal countries have a clear interest in helping tropical states preserve their forests.

One of the biggest ways to help people make better decisions is to demonstrate better alternatives. The New York Botanical Garden had a contract with a major electric utility. When the company built high-power lines, it used to cut a big swatch of forest to keep the trees away from the cables. The botanical garden explored alternatives to this approach. They showed the company how it could operate through the winter without having to worry about the erosion that resulted from clearing the land. This saved the utility money and saved quite a bit of the landscape. When the company puts lines in now, it plants trees underneath them, controlling the height of the trees by chemical or mechanical means. We no longer have the bare earth that was left before.

There are ways of demonstrating new, more sensitive techniques to industry. In a surprising number of instances you can find common ground between economic and environmental interests. Environmentalists must realize that compromise does not mean abandoning our principles. Resource users are not the only ones who have to develop some understanding.

QUESTION. Isn't one of the problems with species protection the fact that it seems so far removed from the needs of everyday people? All the benefits seem very distant, the results of years of research.

ANSWER. Let me tell you about durians and bats. Durians are by far the world's superior fruit. Even tigers eat them. They are the great prize of the Far East! Now, durians are pollinated exclusively by one species of bat. The bats roost gregariously in limestone caves, which are presently being destroyed to make cement. If we can get people to understand that they will have to do without durian fruit if all that limestone is converted to cement, a lot of them will start listening. These bats fly—they commute—up to 45 kilometers each way every evening, and their "gasoline" is durian nectar. Sometimes the benefits are quite immediate.

QUESTION. I am wondering about the cultural and emotional benefits of species preservation that you mentioned. Are they part of "conservation for sustainable development"? Will people in poor parts of the world accept these notions?

ANSWER. There are segments even within IUCN that disagree very strongly with the views of the Species Survival Commission. Some politicians, governments, even biologists, feel very strongly that conservation should always be directly related to man. They feel you cannot convince developing countries to adopt a program if you do not appeal to their economic interests. That is why the subtitle of the strategy is "Living Resource Conservation for Sustainable Development." I am not at all sure they are right.

In my travels I have found people in the developing world who have feelings for preservation stronger than any in the developed countries. The president of Tanzania has said, "I don't care if tourists never step foot in my country again, we are going to preserve our parks and our wildlife because they are part of our heritage." There is no narrow economic "self-interest" in that statement.

Because of our ethical, cultural, and historical traditions in the West, we have a particular perception of nature. We must realize that there are people in other parts of the world who think of nature in completely different ways. These civilizations have philosophies that are much more relevant to the conditions in which they are living than the one we are trying to impose through conservation programs. In this light, timber exploiters with their particular purposes do not seem very different from conservationists. Both are trying to force an entire world view on other people.

There is another approach. Some groups in India, for example, preserve the forests and the mountaintops because these are the places Vishnu lives. Their religious belief is a far more effective way to conserve watershed forests than any police operation. By preserving their sources of water, they get three crops of rice per year. Similar beliefs are common in almost all "primitive" cultures. If we could somehow tie those traditional beliefs to scientific environmental management, our persuasion would be much more effective than any of our present strategies for controlling the misuse of resources. More important, both parties could discover the fundamental similarity between their goals.

Sustainable Use of Species and Ecosystems

GERARDO BUDOWSKI

For the sake of honesty, I should warn you that I will be rather biased, talking about things I know (or believe I know) and leaving aside matters that may be very important but are outside my direct experience. I may also hurt your sensibilities at times by using some words that upset conservationists. I want to present conservation and the sustainable utilization of species and ecosystems as nothing less than powerful tools for development. I know that *development* is a bad word in some places, but it certainly is not, or not yet, in many of the tropical countries that will be the main focus of this chapter. In those countries, if we want to get support for certain types of conservation actions, they must be based on the development of communities, on improvement of rural or city life. Whatever the program chosen or the strategy pursued, it must include the idea of development.

When we speak about sustainable utilization of species and ecosystems, one essential concept must be analyzed, understood, and applied—*carrying capacity*. The idea of carrying capacity implies that ecosystems can tolerate a certain amount of intervention under normal conditions. This can be removal of individuals of a species or other types of exploitation. Beyond that amount of exploitation there is degradation, no more recuperation, and, of course, no more sustainable utilization. I am sure you are all familiar with this basic concept, so I will focus primarily on its application.

Short-Term versus
Long-Term Perspectives

Let us look at resource management from an economic perspective. There are many ways of approaching the economics of sustainable utilization. For instance, Sidney Holt has pointed out that it makes perfect economic sense to exterminate the whales. When you harvest them (a euphemism for cruelly killing them), sell their products, and make a large profit, you then can invest that profit with interest and make more money than if you keep harvesting the whales on a sustained yield basis. From a strictly economic point of view, based on a simple cost-benefit analysis, one may argue that this is the most "efficient" way of dealing with resources.

But if you begin assessing many of the other aspects of the whale slaughter—environmental stress, loss of an irreplaceable resource, loss of educational opportunities, as well as scientific, aesthetic, and ethical values—then the loss can be considerable. Of course many of these values cannot be evaluated easily, but we know that they exist and are worth something. Naturally, these values change according to cultural factors and other variables. Nevertheless, their total is much greater than the strict value of marketable products.

For some people—and they number hundreds of millions—the economics of profit means little, often nothing. Ultimately what is at stake for them is their own survival. I refer to the millions of poor farmers who cut the forest or hunt to extinction to feed their families. They are humans like us, who only want to avoid starvation. How can we convince them to worry about plant or animal species when they themselves must fight for survival? This is just another—albeit terrifying—ethical consideration we have to take into account when we speak about sustainable utilization.

If I were pressed to explain why there is so little concern for sustained use and for the concept of carrying capacity throughout the world today, I would have to blame it on a lack of understanding of fundamental ethics. Of course these ethics need to be adapted to each culture and community, but has this ever been attempted? Quite the contrary,

in many places the ethical standards work against the concept of sustainable utilization. Indifference or reactions such as "Let us make a quick profit . . . now" or "*après moi le déluge*" are extremely widespread (even when not stated quite so bluntly). The destructive fishing by highly sophisticated ships from industrial countries along the coastal waters of many developing nations provides an excellent example. The fishermen are pressured to harvest as quickly, as efficiently, and in the greatest quantities possible because if they do not, someone else will. There is no question of sustained utilization. The "cut-and-get-out" practices of many tropical timber exploitation groups show the same lack of regard for anything other than short-term profit.

Finally, we must consider our priorities as part of human society. Usually we think, Country first, and What can I do for my country? But is this really justified? Why are we not educated to think of planet earth first? Should we not be looking at the earth as the first and most primordial element in our moral obligations? If the answer is yes, then it is easy to transfer the concepts of carrying capacity, and eventually sustainable utilization, to a global scale.

Yet another ethical problem has really been compounded by existing educational systems. Mankind is the custodian of a heritage of a great genetic diversity, but this idea has seldom been incorporated into our ethical education. The same applies to other credos of environmental ethics, such as the promotion of diversity—and I mean not only physical or biological diversity, but cultural diversity as well. Ecological ideas definitely argue against today's tendencies toward cultural homogeneity or, even worse, the open imposition of one culture upon another. They provide a counterforce to the trend toward cultural domination. We know of many cultures that had definite ethical standards concerning the utilization of resources. Most "primitive" cultures hunt only for their direct needs. But their standards have often been broken down by the arrival of "advanced" cultures, which make economic profit a much more powerful motivation. As these new ideas are accepted, former ethical standards are destroyed. We have watched this happen to the American Indians and to other cultures in a variety of places.

Perhaps the most important ethical consideration about natural resources applies to decision making, particularly when it affects our future. When we are choosing among several possibilities for managing resources, we must examine the ethical foundation for our choices. The logical and ethical decision will be the one that favors long-term sustainable utilization. Even more, one could argue that in case of doubt the best decision is the one that keeps options open and transfers decision making to future generations. Do we see these concepts as part of resource management today? No, in fact we see a tendency to the contrary. Our future options are diminished by opening up virgin lands, destroying natural diversity, causing extinctions, and increasing the homogeneity of landscapes. Naturally these are not the avowed reasons for the decisions we make, but they lead to the same results.

Examples from Tropical Rain Forests

Having defined some important problems in the abstract, I would like to present a few case studies that illustrate my concerns. I am most familiar with sustainable utilization of tropical forests, which are complex ecosystems that are being rapidly depleted. Tropical forests are more than simple forests; they constitute a series of systems that have great impact on the lives of many people, both those who live nearby and others who are far removed.

Recently studies have calculated the loss of tropical forests to be between 10 million and 20 million hectares per year, approximately 25 to 50 million acres. Analyses are complicated by discussion of how much area reverts to secondary forest or stays as pasture, and so on. But whatever the data, everyone agrees there is an incredibly high rate of tropical forest loss. Table 5.1 outlines projected tropical rain forest depletion in the regions where rates of destruction are greatest. Its figures are based on a study prepared for the timber industry, so they reflect concern only for that industry's use of the forests and not for other uses (including mainte-

FIG. 5.1. A cloud forest in the Arenal reserve. This watershed supplies Costa Rica's largest dam. At 1,200 meters clouds sift through the trees and coalesce to form droplets that feed local streams and, through them, the whole watershed. Photograph by author.

nance of genetic diversity) or the impacts of logging on those uses. Despite this limited frame of reference, however, the study is valuable as the most recent analysis of tropical rain forest loss and the only one to permit a relatively reliable regional breakdown.

It is quite evident that within less than two generations no more tropical rain forests will exist, except for a few national parks and other protected areas. Obviously, there is no question of sustainable utilization of these forests. They are simply being destroyed—cut down and burnt. What is worse, the utilization following removal of the forest is seldom sustained. Why this senseless destruction?

The problem will not be easy to solve. Tropical forests, I have to confess, are not easy to manage. At present, there is not a single documented case throughout the tropical world—in the Americas, Africa, or Southeast Asia—where the mixed tropi-

TABLE 5.1. *Projected Loss of Priority*
Tropical Forests, 1975–2000

	Total Closed Forests[a]		"Operable" Hardwood Forests[b]	
	area 1,000 (hectares)	% of 1975 area	area 1,000 (hectares)	% of 1975 area
West Africa	6,600	47.1	6,600	54.7
Centrally-planned tropical Asia	6,300	29.1	6,000	35.3
South Asia	16,400	23.0	13,600	27.9
East Africa and islands	3,300	17.8	3,200	50.4
Insular Southeast Asia	21,600	16.5	20,000	26.3
Central America	10,900	13.4	4,600	23.9
Tropical South America	64,200	12.0	57,300	13.3
Continental Southeast Asia	4,100	10.6	4,000	13.3

SOURCE: IUCN, with UNEP and WWF. (1980) *World Conservation Strategy: Living Resources for Sustainable Development.*
Gland, Switzerland: IUCN/WWF; Nairobi: UNEP.

[a]Closed forests include logged-over forests "as long as they are not alienated from non-forestry purposes."

[b]"Operable" forests exclude protected forests (parks, wildlife reserves, and so on); forests on terrain that is too steep or wet
to exploit; and forests, such as most mangroves, "permanently without industrial wood potential."

cal rain forest has been managed for timber production on a sustainable basis for a reasonable amount of time. There are, of course, theories about how this could be done, but no one has implemented them successfully. Although we speak constantly of exploiting the tropical rain forest "rationally," no one has yet been able to achieve sustainable utilization. I should make clear that certain types of tropical forests can, of course, be managed rationally, with a profit, and on a sustained yield. But these are not the immense tracts of mixed primary rain forests.

Moreover, cutting wood from the tropical forests is not the principal cause of their destruction by far. In Latin America millions of hectares are being cut every year to create pastures with low carrying capacity. Grazing is managed on a very land-intensive, low-labor basis. There is often only one head of cattle for every 5 or 10 acres of land. The grass-fed cattle produce lean meat, which is exported, mostly to the United States, European countries, and a few meat-hungry countries in Latin America who can pay the price, such as Venezuela and Mexico. The worst part of this problem is not so much that the forests are converted to pastures but rather that after a few years, in rainy areas particularly,

many of the pastures thus created do not sustain themselves; they eventually revert to secondary brushlands or forests of very little value. In other words, the forest has been destroyed, the grass has been used for cattle for perhaps ten to fifteen years, and the area has been abandoned. Then the process is repeated on the next tract of forest. This is the fate of very large forest surfaces throughout Latin America and, more recently, other tropical areas.

In U.S. inquiries about these practices, it was learned that the lean beef from tropical countries is being mixed with meat from grain-fed beef (which has a high amount of fat) to meet the minimum standards for fat in the hamburger industry. Some of the largest cattle ranches exporting beef from Latin America to the U.S. hamburger industry belong to major U.S. meat-packing corporations.

The wood chip industry, also supplied by the tropical rain forests, is another major ecological threat. Very large companies based in Japan and a few other countries, including the United States, are particularly eager to harvest enormous quantities of chips and reconstruct them into compressed or hard boards. They are willing to pay nearly any price—including "under the table" payments to key decision makers. Chip harvesting can hardly claim to favor sustained yield. In their own countries many companies are extremely good custodians of forest resources, and they eagerly practice sustained yield; they often apply strict conservation measures and take special care to replant trees and promote conservation education. Unfortunately, they are not as anxious to be good land stewards in developing tropical countries.

There have, of course, been other reasons for destruction as well: breaking down of traditional land use practices, adoption of new devastating ways to harvest timber, and use of heavy soil-compacting machinery. In many places we know indigenous communities have been using the forest for centuries, yet it is still relatively intact. New technologies, economic ways of looking at (and exporting) resources, and sometimes destructive governmental interference—such as huge colonization schemes with populations totally alien to local cultures—have brought havoc to both people and their resources. In 'the tropics this new resource utilization has reached and passed critical thresholds beyond which the land loses its capability.

The land is being stripped of its productivity. But at the same time, the new population that I mentioned needs to continue feeding itself, and it can only do this by cutting down more areas of forests, including lands clearly marginal for sustained cropping. Some of the most blatant examples of this are along the Trans-Amazonian Highway. Peasants throughout Latin America have been relocated to new settlements in the forest. They have not come near the planned goal of making a permanent living from the land. Politically, it is extremely expedient to relocate land-hungry farmers. For politicians who are in office only a short time, it is a glorious moment when, with all the news media present, they distribute plots to farmers and award them property titles. For the farmers there is some initial hope, but ultimate disaster. They cannot make a permanent living from these marginal lands, regardless of ownership. The lands usually have poor soils or they are too wet, too dry, or too steep for farming. Let us not be fooled—there are no more large tracts of good agricultural land left unused anywhere in the world. The amount of land

that will produce a sustainable harvest is extremely limited, and it is already in use.

Possible Solutions

Enough lamenting. Let us now search for some solutions. The first thing to consider is the interdependence of all the inhabitants of planet earth. This is not something on which I will elaborate. It should be clear in developed countries such as the United States, which import a very large amount of their resources from the rest of the world. A U.S. citizen uses 500 times more water per day than a citizen from an African country. But the U.S. citizen also uses many times more energy, minerals, and other resources that do not come from his own country.

You may argue that this favors trade; buying raw materials such as timber, chips, or beef from another country helps that country obtain currency to meet its balance of trade. True enough. But will it last? Does this system provide a sustainable yield? And who within the exporting country gets most of the benefit? What is the United States, or any other of the receiving countries, doing to assure—for its own sake and the sake of the exporting countries—that this exchange of resources for money is based on long-term priorities? At present, the ability to sustain trade is simply not a major consideration. It is not really necessary to trace the blame. Neither the importing nor the exporting countries give this matter much thought.

Renewable resources can and should be managed for sustained yield. The richer nations have a stake in assuring that the trade in vital raw materials will be maintained on a permanent basis. But this idea is largely ignored in commercial decision-making processes, assistance policies, and many current banking or financing schemes. At best we see policies of conditional involvement in remedial action. Of course, there are two schools of thought about international involvement: one advocates nonintervention and the other argues for greater efforts. It is not for me to judge when active involvement becomes intervention in the internal affairs of another country. But certainly neither isolation nor business as usual can be justified. Let me say it another way: In a shrinking world—shrinking in terms of the available stocks of natural resources—we cannot afford to project the past indefinitely into the present. Again, the concept of critical thresholds is central. There surely are ways of demonstrating the need for new policies. Unfortunately, however, I do not see governments making any great efforts to reverse present trends.

Foreign assistance and cooperation, the "compensation" to help poorer countries protect their natural ecosystems and resources, provide examples of possible solutions and their difficulties. The Brazilians were very upset some years ago when scientists insisted that they should not cut the forests along most of the Amazon. It is "our country" they claimed, and "it is our sacred right to develop the Amazon as we see fit. We do not accept any interference." But I think that time has passed. The Brazilians themselves now know the problems with Amazonian development. Of course they still do not want interference, but they certainly would like to receive support, particularly in scientific and educational areas. Smaller and poorer countries need such assistance desperately. They need to know how to study, to plan, to understand how their ecosystems function, and to manage their resources. Good management sometimes means protection, in most cases it means wise use. Even sustained yield does not necessarily mean direct exploitation. A national park, for instance, is land that can be managed on a sustained yield. It produces educational, scientific, ethical, and other values as well as products such as water and wildlife. We must also encourage certain aspects of traditional life-styles as ways of managing resources. These traditional systems do not deplete resources. Unfortunately, we have been led to believe that new technological advances can solve all problems.

Some international organizations are promoting this new kind of cooperation, but their work has been too limited. It is unfair to request that a poor, undeveloped country build a representative system of national parks, biosphere reserves, and other protected areas without outside assistance. After all, who are the main users of parks and reserves in tropical countries? More often than not they are people from richer countries and only the wealthier segments of the local population.

There is still another possibility for cooperation between richer and poorer countries: joining forces to promote national conservation strategies. I was in Venezuela some months ago to contribute to a publication entitled "Presentation of the World Conservation Strategy: The Case of Venezuela." This document, based on the World Conservation Strategy, can easily be adapted to other countries and conditions. The implementation of both national and international strategies deserves worldwide support. At present, only some forty of the 160 or so countries have published reasonable conservation strategies for themselves. The effort of unifying the implementation of these strategies with implementation of the world strategy—remember "Only One Earth," the unofficial motto of the 1972 Stockholm conference—remains virtually untried. We must pay much more attention to implementing unified conservation plans.

How can we approach the biggest environmental problems, such as relieving the pressure on tropical forests? There is obviously no simple solution, particularly if the human population is continuously expanding. However, certain rules or guidelines can help.

The first of these guidelines is better utilization of the areas with high potential productivity that have already been opened. In tropical regions, these would be lands that are not forested and already have some infrastructure. It is incredible that places with fertile soils in Colombia, Venezuela, Mexico, or Costa Rica are being used as extensive, low-yield animal husbandry estates that produce beef primarily for export. Even though the yields per unit of surface are low, the landowners make a good profit because of the size of their ranches. With better management techniques the same land could produce ten, perhaps fifty, times as much food. Instead of being a source of beef for export, it could sustain local families. However, this may not favor the interests of the landowner, and under the prevailing political system it is almost impossible to change land use patterns. The landowners have a perfect right to use the land as they do under the present constitution. But something is wrong with a system that allows a person not to make the most of this land on a sustained yield basis when there are scores of hungry people in the same region. Admittedly, this is a simplistic analysis of a complicated problem affecting basic rights. Nevertheless, it is quite clear that if some of the best agricultural land in these countries was properly used, there would be absolutely no need for encroaching on marginal land, which cannot be managed for sustainable yield. Eventually, with population relentlessly growing, some long-term alternatives involving proper carrying capacity must be sought.

Deforestation is closely linked with pressures on land use. Eventually, we must manage all land so that it produces its maximum sustainable yield. Marginal land should be left largely untouched. Yet today almost half of exploited tropical land is marginal and is condemned to be destroyed. When it is lost we will also lose the wild animals and plant species that it harbors. These species themselves would have been the sources of new food and medicines. In some cases pressure on marginal forestlands can be relieved by carefully designed forest plantation programs. Needless to say, these should never be established at the expense of natural forests. But on degraded lands, such as the ones resulting from abandoned pastures, they can be quite beneficial. We urgently need some imaginative investment schemes by capital-rich countries to promote these plantations and to help poorer countries build industry based on their products. They could also supply timber and fuel wood for local needs. I repeat: Plantations should never be started at the expense of the original forest. They should be developed on land that has already been stripped of its original forest cover and is often subject to erosion in its present state.

Trees that fix nitrogen from the air through bacteria associated with their roots could be planted on a large scale and become the basis of a series of industries. They would simultaneously produce charcoal or firewood as well as stop erosion and rehabilitate the soil. This need for firewood is becoming a critical problem in many places. There are also possibilities such as agro-forests, in which trees are interplanted with food crops. All these ideas need to be explored.

FIG. 5.2. *One of many landslides in Costa Rica planted with cuttings of Itabo,* Yucca elephantipes. *This species holds the soil and develops into small trees which produce edible flowers and cuttings that can be exported as ornamentals or used for additional plantings. Photograph by author.*

If we want to institute sustained yield, we must promote long-term conservation policies for land use and management. There must be guidelines for decision makers based on scientific, ethical, and other cultural considerations. Of course, there is also a place for economic analysis, but it must be carefully weighed for short- and long-term goals and include all the direct and indirect, or intangible, benefits. And there must be an underlying sensitivity to basic ecological concepts. Political and economic land use decisions must be based on ecological considerations, not the other way around. We must strengthen sensible technical cooperation programs, be they multilateral, bilateral, governmental, or nongovernmental. I realize that the meaning of the word *sensible* can be anyone's guess, but these programs should not be based strictly on the sponsor's political or economic needs. They

FIG. 5.3. *A live fence of Guachipilin,* Diphysa robini-
oides, *in Costa Rica. The wood of these trees is valuable
for building, their roots fix nitrogen, loppings produce biomass for soil enrichment, and the branches are good fuel.
Scientists, however, know practically nothing about the
ways local landowners cultivate it. Photograph by author.*

must pay much more attention to sustainable yield,
carrying capacity, long-term objectives, and, above
all, building of local leadership. It is also essential
to strengthen educational programs and local research facilities. The conservation of natural areas
in developing countries should be a genuine worldwide concern. These areas promote the welfare not
only of our towns, our provinces, and our countries, but simply of planet earth, our home.

Discussion

QUESTION. Do the methods of cultivation you are
studying expand on local practices or make basic
changes?

ANSWER. For the most part, they expand on local
practices. For example, the leguminous tree Gau-
chipilin is extremely fast growing and produces a lot
of nitrogen. Local people plant it because they say
it makes the soil "more fatty"—it increases the fertility of the soil. The interesting thing is that these

local farmers have been using it for years to improve their soil. Our duty as scientists was to understand why they planted it and how the system
works. After studying their methods, I can say that
they work very well indeed. This is one of the trees
that has been brought to Thailand as a replacement
for the poppy cultures to make the soil productive
for sustained yield farming.

QUESTION. Are the methods that you described
solely an approach to helping subsistence farmers—
which is a valuable enterprise in itself—or are they
likely to become even more widespread?

ANSWER. There will be fundamental changes in
agricultural methods in the future. What they will

be is anyone's guess. We need hard thinking about how agriculture will evolve in the next years. But one thing is certain; agriculture will not be what it is now—high input, high everything. You cannot long afford to live if you use 10 to 15 calories of energy to produce 1 calorie of food. It will simply not be possible. The Chinese farmers are producing 1 calorie of rice with 0.05 calorie of input. American farmers do it the other way around. Something will crack under the strain of such energy use. The agriculture of the future will favor solutions that save as much energy as possible.

The concept of carrying capacity is also one we should dwell on. Ultimately, the carrying capacity of the land around us will determine how long we survive as a species. There are indications that we have already exceeded our carrying capacity, or at least that we are on the edge of it. We can get away with this for a while because the results will not manifest themselves right away. But we will pay the price if the human population continues growing.

Contrast this approach with the economic one. There is a theory which states that in the long run the dollar will prevail over all else. The pendulum of public opinion sometimes swings toward protecting natural resources at all costs; at others it swings back toward the dollar. Right now it is swinging back to short-term economies. But we will have to think in terms of long term survival as a society.

QUESTION. How have you been able to motivate people to accept these new ideas? Do you use government policy, education, or some other strategy? It seems to me that the hardest step is not discovering better techniques, but getting people to adopt them.

ANSWER. We capitalize on the experience of other countries. Brazil has plantations almost half the size of Massachusetts. They have one of the largest wood and pulp exporting industries in the world based strictly on these plantations. How did Brazil do it? They have a law that allows landowners a tax deduction of up to $1,200 per hectare of disturbed land that has been planted again. These tax incentives have also been used in other countries. Another stimulus comes from an American company, the Scott Paper Company. It gives farmers nursery plants and assists them with cultivation in the understanding that they must sell the crop to Scott. The cropping is done after eight to ten years and sold at the normal price. There is a common inter-

est here between the company and the small farmers.

With the eucalyptus, it was also the small farmers who saw a chance to better their condition. A local farmer began planting eucalyptus not for economic reasons, but because he liked trees. As a result he is now making money. He does not quite believe it yet. A new idea must be demonstrated successfully to get others interested. That farmer is making money and others are asking how.

The basic idea of live fence posts is very old. They have been planted for hundreds of years but never investigated by scientists. It was always considered a poor man's fancy: Because they could not afford stone or some other material they stuck in live branches. Now suddenly we are discovering live fence posts. I have the suspicion that the few scientists who knew about live fences never really believed that farmers could develop such a good idea by themselves. But we have come a long way. We now feel that the farmers have much to teach us. Rather than transferring techniques from outside and having the farmers refuse to adopt them, we begin with their own systems, make improvements, and demonstrate the benefits. They can see the results for themselves.

QUESTION. What percentage of farms with multicropping are operated by small farmers?

ANSWER. The answer depends on the crop. Ninety-nine percent of all coffee and 100% of cocoa is part of a multicrop system. The remaining 1% of coffee is grown by very rich farmers who were very impressed with the high yields of monocultural stands before they understood that you have to make extremely high investments to maintain them.

Farmers, and not just the wealthy ones, use the multicropping system for many reasons: It is a protective device against excessive rainfall, and it also brings nutrients up from the deeper soil. If you apply fertilizer in a multicropping system, the fertilizer goes a long way, not only to the crops but also to the trees, which bring it right back to the surface of the soil. Another major factor is psychological. No farmer wants to depend on one crop. He needs something else to fall back on. If a rich farmer loses all his crop, he goes to the bank and gets a loan. The poor farmer cannot do this, so he develops a system that has many component crops. Then, even if one fails, he does not perish.

We are just discovering these reasons for multi-cropping. We have been misled by technologies transferred from other countries, which are not adaptable to the ecological and socioeconomic conditions of the local inhabitants. Our whole strategy now is first to learn how the local farmers operate. Then we see how the system can be improved.

QUESTION. There are large corporations and small landlords buying property, cutting down the trees, sawing them, and then leaving the land after it is depleted. It seems to me that countries where this is happening have to understand what is going on and legislate against it. You say countries need our help to do this. How can we help?

ANSWER. First, we must help them understand the issue. Schools are poorly staffed and equipped and people cannot travel. Rather than doing the work directly, we must help these countries develop the skills they need.

You can lobby the decision makers in your country. If there is more awareness in each country about the process of decision making, many of the people at the tops of governments will not be able to get away with the mistakes they have been making. For example, we need help to build up initial money and expertise so that local conservation organizations can have access to small libraries and consultants on issues such as deforestation. Very few local people know much about how to prevent this problem. They have never had the chance to be told, to be given examples of the sorts of problems that can arise.

Another thing we can do is relieve the pressure that causes deforestation. For example, the cattle industry is promoting deforestation in a criminal way. This is not because of the nature of the industry itself, but rather the way it is practiced now. It would be possible to breed cattle on a sustained yield basis, using some of the areas that are fully managed. It would also be possible to create forests that can maintain themselves while producing a sustained yield of timber. We can help countries strengthen themselves and at the same time relieve the pressure on the natural resources demanded by developed countries that leads to destruction.

COMMENT. When I was listening to you talk about the destruction of tropical rain forests, I thought about the resources we have here in Massachusetts

that we are "mining" the same way—throwing away rather than husbanding. There are two main ones, and there has been a different resolution to the management of each.

The first are the salt marshes on the coast. We have about 40,000 acres of these left. They are the remnants of a system that was much larger at one time. We now have some fairly protective legislation that keeps people from altering and dredging them. But the legislation has not been there for very long, and the coastal management program of which it is a part is really the remainder of a grand scheme that fell apart—we tried to develop some broad land use legislation in the United States. The only portion that did pass Congress was the coastal part. I think that happened because there was a lot of good information being generated at our universities on salt marshes and coastal land.

The other resource is agricultural land. It is coming to be seen as a more valuable resource as increasing fuel prices make us realize that importing a large amount of our food makes us very vulnerable to rising transportation costs. We are trying to develop a protective system for our agricultural land. But it has been very difficult to do this because we have to change people's perceptions of the relative values of different land uses.

III Priorities for National Action

National and Regional Conservation Strategies

KENTON R. MILLER

Achieving the goals of conservation means that each nation or local unit of government must take action to focus its activities on priority issues and the removal of obstacles to them. Procedures and principles for strategic planning can help those involved to work as expeditiously as possible. In contrast to traditional approaches, strategic environmental planning requires a careful analysis at the early, conceptual stage of a project to avoid irreversible or costly commitments. Such global thinking often reveals that the very laws and institutions which govern natural resources are often parts of the problem rather than instruments for the solution.

Strategy Formulation

In formulating strategies it is possible and necessary to (1) determine the priority requirements for achieving objectives; (2) identify the potential obstacles to meeting these objectives; and (3) propose cost-effective solutions to overcoming these obstacles.

Four principles are useful guides to strategic planning:

1. Integrate the various elements involved. Laws, management agencies, use practices, zoning codes, and other factors are all part and parcel of the same exercise. It is imperative to avoid separating aspects of the problem simply by following existing structures or traditional ways of doing things. We must categorize according to the way natural resources

function and their relationship to society's requirements.

2. Keep all options open to retain future flexibility. Our knowledge of natural resources and society is limited. To set plans and solutions "in concrete" suggests a great deal of arrogance and will leave future generations with few choices about how to use their natural resources.

3. Mix cure and prevention in an effort to deal with current pressing problems and to reduce the hazards of present-day activities that will affect tomorrow's people. By carefully thinking through the environmental problem, it is often possible to blend short-term problem-solving measures with long-term investments that avoid future crises.

4. Focus on causes as well as symptoms to get at the root of the problem and to aid in understanding the situation as a whole. For example, riverbank stabilization without upstream reforestation may be a wasted effort, like using a Band-Aid to cover a cancer symptom. Strategic thinking involves constant vigilance in observing and responding to symptoms that indicate coming problems. Symptoms require immediate corrective measures and, at the same time, action to cure the root problem.

These basic ideas can be seen operating in present natural conservation areas. The nations of the world have established over 2,500 national parks and other types of protected areas, totaling some 400 million hectares (IUCN, 1982). Traditional thinking considers this is a lot of land and water dedicated to protecting wild species. It is often suggested that we cannot afford to set aside more land for parks and reserves because of our growing requirements for food, housing, and other human needs. To analyze this suggestion strategically, other questions need to be examined first. For example, one objective of conservation is the maintenance of genetic diversity (Soule and Wilcox, 1980). In order to know that the right land has been set aside as parks, we have to ask if established conservation areas were selected to include the millions of species and their diverse forms. Current study suggests that there are fundamental problems

with present efforts to protect diversity. Returning to the above-mentioned principles, these problems include the following five considerations:

1. Existing areas were generally chosen to meet other objectives, including preservation of spectacular scenery, recreation opportunities, and outstanding natural formations—all of which are valid and important. However, these marvelous natural areas do not always maintain genetic resources.

2. Conservation areas have usually been set up in isolation from regional development plans. They are typically envisioned as "islands" of nondevelopment (or antidevelopment). This lack of integrative planning has obscured the key linkages between natural areas and water conservation, seed sources for agricultural development, erosion control, research on wild species, and the study of geologic phenomena such as volcanism.

3. Most parks are being surrounded by farms, pastures, or even residential tracts and mines. Flexibility for realigning parks to relate to the new objective of maintaining genetic resources are rapidly being reduced, and for this reason society's options for new pharmaceuticals, foods, and industrial chemicals are also being diminished.

4. Approaches to setting up parks have tended to be site specific, often geared toward a site to be saved from imminent development. On the one hand, different opportunities that better serve society are often missed, and, on the other hand, the values that planners did try to preserve are not studied sufficiently to offer a strategy that will work over the long run. For example, through careful placement of a conservation area, it is often possible to provide protection for watersheds at little additional cost. But haste has often led to the creation of a conservation area that does not meet the biological needs of the very species the area was created to protect.

5. Finally, the concern for dealing with symptoms—the decline in numbers of a particular species—has obscured the *cause* of those symptoms—the destruction of habitat. Although action to pre-

vent the loss of individuals is useful, failure to ex- amine the causes of habitat loss can lead to a spe- cies' extinction. Preventing habitat loss may be difficult when its cause is distant and seemingly un- related to the problem, such as when changing markets for agricultural commodities shift land uses (Myers, 1979).

Obstacles

Conservation objectives are difficult to achieve be- cause of a series of obstacles, six of which are of par- ticular concern. Continuing with examples from the problem of protected natural areas, let us look at the sort of problems caused by each of them.

1. Conservation is usually excluded from policy formulation. At the policymaking level, where consideration is given to agriculture, forestry, and water resources, conservation concerns are seldom involved. Thus, the potential of national parks, forests, refuges, and other reserves as tools to aid development is often overlooked. One park can protect a water catchment upstream for down- stream irrigation works and at the same time provide in situ preservation of forest trees. Further- more, if these opportunities to support develop- ment while maintaining living resources are missed, elaborate and expensive projects to reforest watersheds, prevent streams from flooding, or res- cue vanishing species are often required later as cor- rective measures. Conservation is not a sector unto itself; it must be seen as an element in all other sec- tors.

2. Laws are often stated and organizations struc- tured in ways that impede conservation. Laws on natural resource protection and use deal with spe- cific issues—water, soil, forest—rather than with an integrative perspective. Thus, laws on nature conservation typically focus on establishing areas in geographic space with little regard for ecosystems and lands or waters adjacent to the conservation areas. This can allow development at the borders of reserves that is destructive to the resources being protected. Similarly, institutions established to manage parks and reserves often have minimal communication with other groups working in adja-

cent lands. This blocks the opportunity to form collaborative goals or, worse, it increases the chances of conflicting activities. For example, it becomes difficult to deal with species that migrate from parks to adjacent lands and with water or air pollution coming from distant areas.

3. Lack of training and basic information thwart professional development for resource planning, management, and general decision making. The various professionals who work in conservation often do not receive academic or other special training. This obstacle involves a lack of training not only in strategic thinking and methods, but in conservation itself. Again, training is dealt with by sectors. One studies wildlife ecology, wildlife man- agement, forestry, law, resource policy, resource economics, outdoor recreation, or range manage- ment. Few schools offer opportunities for integra- tive studies in conservation. Most training pro- grams for park personnel focus on specific aspects of recreation, biological sciences, or environmental communications. Few integrate these skills to pro- vide an overall management perspective. This ob- stacle is even more difficult to overcome because of the lack of information to guide decisions. It is dif- ficult to provide training courses when so few real examples that reinforce conservation theory have been documented. Conservationists are hard- pressed to defend themselves without numbers and facts—"hard data." Political leaders are unable to move forward with innovative legislation and pol- icy if they are not provided with persuasive argu- ments that will hold up in heated debates. Thus, specific claims that national parks can preserve spe- cies and simultaneously serve the economy may be true but are nearly impossible to prove.

4. Decisions to allocate natural resources are sel- dom influenced by environmental considerations. Forests are cleared for agriculture; good soils are covered by concrete for transportation and housing development; streams are diverted and dammed; and wild habitats are altered on the basis of criteria concerning immediate human needs, current mar- ket values, and engineering and design practices. Immediate needs are important criteria for policy, but the consideration of environmental planning

could help formulate decisions that lead to more sustainable development. In terms of protected areas, environmental planning would ensure that areas critical to sustainable development—water catchments and genetic reserves—would be incorporated into the overall regional plan. Furthermore, integrative planning would recognize that protected areas alone cannot pretend to guarantee the maintenance of genetic resources. Even with an optimistic 10% of our landscape under protected area status, the other 90% must be employed in ways harmonious with ecosystem maintenance.

5. Conservation often lacks the support of political leaders, scientists, teachers, and the general public. This lack of support inhibits full-scale action on world conservation and development. The causes are varied but are generally based on a lack of comprehension of conservation's important role in human welfare. Plainly and simply, conservation has all too often been viewed as the hobby of the elite, a way to use leisure time and money, and a duty to keep the nation's heritage. The fundamental issues have seldom been explored openly. For example, in our parks we try to build support by helping visitors enjoy and learn about nature. Perhaps it is time to inform people about the role protected areas play in preserving their sources of water, food, and medicines.

6. Rural development is not designed and implemented with due regard for the principles of conservation. Finally, rural development is among the most dramatic problems facing the world. It is in the rural lands that food is produced, timber harvested, and water captured; these areas are also where poverty abounds, erosion and deserts expand, and species are continually lost. Long-term human well-being requires the integration of conservation and development principles to establish sustainable methods for meeting human needs while retaining the natural resource base.

Now, let us look at the first two of these obstacles in detail. The others are examined in subsequent chapters.

POLICY FORMULATION. Policy formulation is a key obstacle to conservation. It is the point where our philosophy and knowledge come together to guide the choices being made. Thus, conservation must become part of the way we think and a component of our information base. Only then will it cease to be treated as a separate sector, on a project-by-project basis, and as an item of relevance only after an environmental problem arises.

At the heart of the policy problem is the lack of fundamental regard for maintenance. It can be said that conservation is to development what maintenance is to construction. We have shown our capacity to build, yet our efforts to keep our newly formed capital are not always successful. Buildings and roads deteriorate and machines fall apart. Our natural resources face a similar demise. Forests degenerate, soils wash away, nutrients leach, irrigated lands turn to hardpan, wild populations decline, and species diversity is reduced.

Environmental policy formulation integrates the maintenance of natural resources with the other aspects of development. Each new proposal for a road, housing program, agricultural expansion, or national park must address maintenance requirements. Initial design should be influenced by anticipation of the need for maintenance. Savings would thus be realized by avoiding the need for subsequent corrective measures. Similarly, conservation can anticipate the needs of development. A well-studied national park proposal can address requirements for recreation, water protection, genetic resource conservation, research, education, and rural employment, among other social needs.

Strategic environmental policy formulation is a serious attempt to link the elements that are parts of particular problems, regions, or ecosystems. Because it appreciates both natural and social systems, it acknowledges both resources and people as integral parts of each exercise. For example, entire river basins are seen as regions for planning. Policies take into account the ecological ties among upstream forestry practices, downstream agriculture, urban and industrial development, and nearby offshore fisheries. Forestry and fishery policies are realized to be inseparable. Parks, contour plowing, seed banks, research on fish populations, and air quality

FIG. 6.1. *New settlements being established following slash and burn operations on the outskirts of the Dumoga-Bone National Park, Indonesia. Photograph by Michèle Dépraz from the World Wildlife Fund.*

monitoring are all parts of one overall regional program.

Ambitious? Realistic? The record of resource destruction, loss of human life caused by flood and crop failure, desertification, and other indicators of environmental imbalance suggests that we cannot afford to continue with sectorial development, the separation of conservation and development, or compartmentalized thinking.

LEGISLATION AND ORGANIZATION. The second major obstacle to sound living resource management is the lack of effective legislation and organization. Mandates for action in governmental policies need to be in harmony with the objectives and principles of conservation.

Laws dealing with natural resources and the environment often ignore the way natural resource systems function—for instance, when fish swim they do not respect territorial boundaries; streams meander; habitats evolve. In the same way, laws disregard the needs of people—subsistence hunting and fishing, relief from their desperate poverty, and facilities for learning new land use practices. Thus, many laws are bound to be neglected because of the inconsistency between their provisions and people's needs.

Organizations are needed to implement programs within the framework provided by new laws and policies. It is common for several agencies to share an overlapping mandate, or for no agency to be held responsible for a particular resource. Alternatively, fragmentation often causes forests, fisheries, wildlife, natural areas, and water to be managed by separate institutions. Even with an integrative view of conservation and development, little can be accomplished until the bureaucratic obstacles to implementation are removed.

Several suggestions can be offered to solve the conflict between conservation and development.

The primary legal structure of each nation, be it a constitution or other legal instrument, should be examined to ensure that it includes a commitment to conservation. Such a commitment would include reference to the stewardship role of the state—its duty to maintain living resources. It would also address the duties and rights of citizens concerning the environment. Under this umbrella, specific laws should be designed to accomplish conservation objectives.

Returning to the example of protected natural areas, a legal structure would set the stage for establishing a network of parks and reserves, which would be designed to meet a specified set of objectives. The government would attempt to preserve genetic heritage, water, scenery, and other resources through areas under protective management. The system would also include areas set aside for wood supplies, fisheries, soils, and other needs.

Then the laws would need to be enforced. Presuming that the laws were ecologically, economically, and politically sound, they would have to be administered by adequately trained and funded personnel. A well-designed set of nature reserves and timber production areas will not meet social and economic commitments unless people are taught to respect the areas—and taught that they will be held responsible if they do not.

Organizations responsible for natural resources must be expected to provide stewardship for the nation's resource base. Seen in the context of sustainable development, this is a very important mission. Governments must examine these organizations to ensure that their legal mandates, funding, and personnel are appropriate to this responsibility. It is interesting to note that most nations commit their national parks "in perpetuity." Yet the limited support they provide to the responsible agencies suggests less than full understanding of the significance of that commitment. Finally, mechanisms are required to provide coordination among agencies within the government. A unity of perspective must guide practical application. Potential impact of each action affecting the environment should be considered.

Various examples of national and international coordination exist to point the way. Several countries have established ministries of the environment, environmental protection agencies, boards, and other bodies. The draft national conservation strategies for Thailand (IUCN, 1979), New Zealand (Nature Conservation Council, 1981), and Australia (Commonwealth of Australia, 1982) offer ways to project visions of unity and means to achieve goals regarding specific problem areas. In each case a network of protected areas and a mechanism for coordinating management agencies is presented. However, there are too few countries which have taken such a strategic view of their conservation needs.

Discussion

QUESTION. Is there room for local knowledge in strategic planning? I am not just asking about input from scientists or engineers. What about fishermen and farmers in the area or others who know the local conditions well?

ANSWER. Local knowledge is a critical factor. I once worked on a project in the Caribbean that required seventeen separate planning maps per island, each one showing a different variable. These maps were all drawn from information given by local people—fishermen; farmers; a man living in the forest, who probably had barely been to school but who knew exactly what we were asking of him and what we were trying to do. The local fishermen made the first draft of the map of coral reefs. We showed it to some marine scientists who said they would have to spend months confirming the data. Remember, these fishermen drew it in a couple of evenings. They already knew which fish inhabited which reefs. We often had to look up the name of a particular species in illustrated guides because the fish have different local names on each island, but the fishermen's information proved to be very sound.

Participation made the plan important to the local people. In the past, developers simply arrived with bulldozers. The islanders' impression of planning and development had been a truck and bulldozer in their front lot, or the whine of a chain saw out in the forest. By actively involving them, we helped make the plan theirs, not ours, and it became important to them to be sure it worked.

QUESTION. Have you ever been approached to participate in development planning in a developing country and elected not to because you questioned the goals or objectives of the political leadership?

ANSWER. There have been some very awkward times. I worked in Chile for a number of years. It was fascinating to experience three regimes and to see what happens in resource management as the political order changes. Actually, I found very little difference in resource policies among the three governments. There was a shift from left to right, but at the level of action on the ground, very few changes were required.

One year I had to switch my mind from working under a communist Cuban government to the rightest regime in Paraguay. My feeling has always been that we are working for the welfare of people, whatever the politics. Making the biosphere operate sustainably is probably the most apolitical endeavor possible. I never felt compromised because I was never in the service of a particular ideology. The IUCN is a special institution because it deals worldwide with resources of great value while avoiding political quarrels. Right-left, north-south, and east-west, all sitting at the same table. When we met in the Soviet Union a few years ago, the South African, Israeli, and Chilean delegates got visas. The environmental issue is important to people in a way that is suprapolitical.

It is extraordinary how many Third World countries have grabbed hold of the World Conservation Strategy and taken an interest in it. The press abstracts are really quite startling. I find it enormously encouraging that the strategy is now in Russian, Chinese, Portuguese, French, Spanish, Arabic, and German. Countries have sponsored their own translations and republications so that the strategy can be useful to as many of their people as possible.

QUESTION. It seems to me that a model for strategic planning is not transferable. You say that the public, regardless of politics, right or left, consistently favors conservation. But doesn't the planning model have to change as you consider different political systems?

ANSWER. I think it depends on how you want to define the word *model*. If you mean the steps you would follow in decision making, the same model should apply anywhere. However, if you mean the details of the political process, there are differences. One country may put certain kinds of resources into public ownership, while another may pass pro-

tective legislation. In this country we have a vast bank of social real estate. We have effectively nationalized large portions of the country—but we do not use the word *nationalize* to describe our actions. We think of public ownership differently. If you went to East Germany, where most of the food is produced on private land, you would never realize how much government control of land allocation there is. The political models vary. There are different local ways to manage land and land tenure, but if we look past the details, there is a basic core of decision-making structure that is generally applicable.

With a little imagination it is usually possible to find a model that works, given the constraints of the public administration system. I have not done any great amount of work in developed countries, but I have a fair amount of experience in developing ones. They have a great advantage over developed countries in their very short time horizons. It is rather easy for them to take a strategic approach, because they do not have long histories of vested interests to overcome. In a country that has hundreds of years' history of institutionalized methods, there is usually strong resistance to trying something new. The developing countries have the advantage of starting without that.

QUESTION. I would like to follow up on that point. Is it even possible to consider strategic planning in areas with a long history of development? Isn't the basic pattern for land use already so well established that it cannot be changed?

ANSWER. In the town of Lincoln, Massachusetts, a very interesting approach to land development has evolved. Initially it was not a grand strategy, but it has taken on the form of a strategic plan. The town has been very successful in identifying areas of conservation concerns, such as wetlands, aquifers, important farmlands, prime forests, wildlife protection areas, vistas and trail connectors, and a variety of things that are important to the community in their natural state. These areas are protected while the town is developing. Equally important, the planners have formulated a strategy for guiding growth in the parts of town that are not as important for preservation. All this is financed through a system in which new development both pays for itself and supports the conservation that is an essential adjunct to that development. It is important to point out that the town recognizes the importance of intangible as well as tangible items, such as the farmlands and wetlands. It is also worth noting that development has included housing for low- and moderate-income families as well as the elderly.

This model is often criticized because Lincoln is a wealthy, well-educated community. There is a feeling that its experience does not necessarily apply to other areas. But although not everything may be applicable, the attitude and the general principles are easily transferable. All governments should identify very specifically their areas of conservation interests that warrant protection and their areas that can be developed. These, together with an examination of where the pressures to develop will come, are the first steps toward anticipating where intervention will be necessary.

Lincoln has been able to provide fair compensation for changes in land value by using a variety of techniques. The town often purchases land and sets it aside so that it will be available when development is necessary. It has changed some of its zoning laws from the traditional 2-acres-per-house standard to allow high-density and cluster housing. This has brought income to the town and kept service costs low. At the same time, this strategy allows much more open space preservation.

One of the most important things to come out of this new approach is that people have stopped reacting defensively to development projects and begun to look at how their town can be developed creatively. There is an increased sensitivity to economic, social, and ecological needs.

In many other parts of the country the desire for economic growth is seen as being in conflict with careful planning. This sends us back to the old haphazard approach to development. There is an implicit assumption that resources are limited. But as resources become less abundant, the need for planning becomes more important.

QUESTION. Many countries' economic systems and incentives tie them to strategies that do not reflect a conservation ethic. How long can you keep strategic planning from becoming political? Is it possible to avoid legislative committees and political bargaining?

ANSWER. Everything is political, everything is economic, everything gets back to ecology. A development package cannot be divided by sectors. There are moments when we put aside our own particular ideologies because our common goals hold us together long enough for a discussion, maybe even long enough to prepare a strategy. There are issues that bind us together. For example, the United States and the USSR have done some rather exciting joint work on fisheries planning. It was very quiet; you hardly heard about it. The Baltic Sea has been governed by written documents that go back hundreds of years, signed by nations that have been enemies for just as long. The real work gets done in these few moments.

Environmental Planning and Rational Use

PETER JACOBS

Traditional conservation action tended to focus on the creation of protected natural areas—national parks, nature reserves, marine parks, or other equivalent areas. Notwithstanding the enormous energy devoted to this strategy, the areas protected represent a very minor percentage of the terrestrial surface of the globe. In Canada, for example, the national parks system accounts for approximately 1.3% of the territory, or 50,060 square miles. Furthermore, the concept of nature reserves related to marine life is still relatively undeveloped, and the total aquatic *and* terrestrial surface of the globe subject to special conservation protection is very small indeed. It must be noted, however, that many of the truly unique and rare natural resources of Canada and of many other countries are protected within these small areas, and that there is an important distinction between the raw quantity of protected territory and its quality.

The initial impetus for the urban and national parks movement in nineteenth-century North America developed from a perceived social necessity to counteract the congested city and to conserve outlying natural wilderness areas already subject to widespread development pressure.

During the first half of the twentieth century, an articulate conservation ethic and growing demand for outdoor recreation stimulated the purchase of vast areas of land well beyond the urban fringe for national, provincial, and regional parks. Toward the later half of this century, the urban population began to colonize the countryside with summer cottages, then winterized residences, and, more recently, permanent residences from which they commute to the peripheries of the cities where more and more jobs are located.

The conceptual polarity between conservation and development, between national parks and nature reserves and the urban realm, between the "greenfix" and the "gray plague" is only gradually being replaced by the more realistic understanding that conservation and development are not necessarily antithetical. Clearly, the conservation ethic must extend beyond the borders of our ecological reserves, national parks, and wilderness areas to embrace the totality of human settlement. Frederick Law Olmsted's principle of "a balance of nature" in the late nineteenth century, which created the urban park movement in America, is still valid. The scale of the urban realm, however, has expanded to include the totality of Canada as it has most of the world's nation-states.

The conservation challenge, in Canada as elsewhere, is to establish a new balance between man and nature. To achieve this balance, the World Conservation Strategy suggests that maintenance of ecological processes, preservation of genetic resources, and the principle of sustainable use must be achieved in developing as well as developed countries. Environmental planning is one of the critical means of reaching these goals. The brief examples of conservation and development issues that follow are designed to explore the relationship and relevance of the environmental planning process to the objectives of the World Conservation Strategy.

World Conservation Strategy Goals

MAINTENANCE OF ECOLOGICAL PROCESSES. Maintenance of the widest possible array of ecological processes is essential if we are to respond to the growing world demand for food, continue development of medicines necessary to sustain health, and respond to the need for shelter, which in many areas of the world depends directly on a continuing supply of natural materials.

Large areas of prime-quality land are being permanently taken out of agricultural use by urbanization. In developed countries at least 3,000 square kilometers of prime agricultural land are submerged

every year under urban sprawl, and in the United States alone 10,000 square kilometers of arable land are usurped each year by industry and urbanization. In addition, close to one-third of the world's arable land will be destroyed in the next twenty years if current rates of land degradation continue (IUCN, 1980).

In Thailand and much of Southeast Asia, in Latin America and Africa, population pressures on forested regions are destroying a traditional and balanced agricultural practice with devastating results for a wide range of ecological processes. In the post-industrial developed countries, many project proposals are so vast and untried that we must rely on "best professional judgment" and the scientific "state of the art" to weigh critical and frequently irreversible alternatives. Evaluation is an important aspect of any planning process in which the maintenance of renewable resources is an essential criterion.

In Canada's Arctic region, assumptions about new and powerful technologies could produce equally disruptive results. For example, an Arctic pilot project has been formulated by Petro Canada to test, on a reduced scale, many hypotheses about the possible and probable impacts of even larger scale petroleum extraction from the Canadian Arctic archipelago. Damage to ecological processes has been approved on the grounds that it falls within reasonable limits and that the very nature of the pilot project concept is to test the validity of the assumptions outlined in the environmental impact statement. But what if some of these assumptions about new technologies, generated in good faith, are marginally incorrect or even wrong?

To achieve the objectives of this Arctic pilot project, two tankers, 140,000 metric tons each, must be capable of sailing through approximately 10 feet of solid ice in the High Arctic. Engines generating 150,000 horsepower will be required—this is ten times greater than the horsepower of Canada's largest Coast Guard icebreaker, the *Louis S. St-Laurent* (Government of Canada, 1980a). Needless to say, little is known about the real noise impact of such tanker engines.

In an initial but carefully argued analysis of the assumptions leading to the conclusion that "noise from ships' engines could frighten nearby animals"

FIG. 7.1. *The metropolitan region of Montreal, population 2.5 million, is situated over the finest agricultural soils of the region. Photograph from Photosur, Inc.*

but that "cumulative impacts on birds, ringed seals, caribou, muskoxen, and whales will be minor," a Danish scientist postulates that the noise generated will raise ambient noise levels by 40 decibels at a distance of 100 kilometers (Mohl, 1980). The range for a given mammal call is reduced by a factor of ten for every 20-decibel increase in ambient noise. In theory, mammals will attempt to escape the noise pollution, but their only means of long-range communication and orientation will be overpowered by the very noise they are attempting to avoid.

The scenario is even further aggravated by the fact that "part of the planned LNG (liquified natural gas) route, with an average of 2 trips per day, may pass in close proximity to the primary wintering quarters for narwhales, belugas, ringed seals, walruses and the few remaining bow head whales"

(Government of Canada, 1980b). It is hard to conclude that a ship generating noise worse than that of a roaring gale passing close to the wintering grounds of marine mammals or competing with these mammals in some of the spatially limited Arctic straits will be "minor."

At issue are two sets of assumptions, both generated in the absence of real knowledge about what is essentially an unknown phenomenon. In one instance, the disruption of ecological processes is assumed to be relatively minor; in the other, the consequences may be dramatic.

PRESERVATION OF GENETIC DIVERSITY. The preservation of genetic diversity is directly related to the maintenance of ecological processes. Failure to manage ecological processes results in the inevitable loss of genetic material, which is both our insurance policy and investment portfolio for the future. Pressures on forest ecosystems such as that in Thailand or potential disruption of marine wildlife in the Canadian Arctic have direct bearing on increasing losses of the world's genetic resources. Notwithstanding our attempts to model and evaluate the impacts of our development proposals, we have acquired significantly greater power to modify the environment than we have to control it. Further, we lack the knowledge base necessary to account truly for the genetic losses that occur.

Species and varieties of plants and animals are being destroyed faster than they are being discovered. More than 1,000 vertebrate species and subspecies and an estimated 25,000 plants species are threatened with extinction (IUCN, 1980). These figures do not account for the inevitable losses of small vertebrate species—let alone of invertebrates—whose habitats are being entirely eliminated.

Consider the following situation in India. Until relatively recent times the bamboo of the forests of the Western Ghats was considered a weed rather than a resource. The advent of paper mills altered this perspective, and industrial use of bamboo has increased by an order of magnitude over traditional uses.

When the first paper plant was established in 1937, the sustainable yield of bamboo was estimated at over 100,000 metric tons. Only a fraction of this potential yield was presumed to be required. Because of the periodic gregarious flowering of bamboos, however, the stock flowered and died in 1954–55, precisely when the paper mill increased its production. In 1958 a second mill was established on the basis of an estimated sustainable yield of 150,000 metric tons. Gregarious flowering occurred in 1959, and following the regeneration of the bamboo stock the annual yield averaged only 40,000 metric tons. A third paper mill anticipated an annual supply of 140,000 metric tons, but competition from the initial mill and obvious overestimates of the region's sustainable yield potential has reduced supply to this mill to less than 25,000 metric tons.

Nonsustainable harvest of the bamboo has important but frequently unrecognized effects on the rural populations in the forests themselves.

Bamboo shoots are an important item of food; bamboo culms serve as the major raw material for house construction and many implements and containers, and bamboo leaves serve as major fodder for livestock in the dry season. Many development projects all over India, particularly paper mills and hydro projects, have overexploited and destroyed vast tracts of bamboo forests. There is, however, almost no information available on how this has affected the quality of life of the indigenous population.

In the absence of such information, and of any awareness of the contribution of wild plants and animals to the quality of indigenous population, development projects are justified in terms of increases in the employment opportunities to these people. However, experience in many parts of the country has shown that the adverse effects of the depletion of bamboo stocks more than counterbalance the positive effects of increased earnings (Gadgil and Prasao, 1978).

SUSTAINABLE UTILIZATION OF RENEWABLE RESOURCES. Sustainable utilization of renewable resources is an obvious consequence of conservation concern over the maintenance of ecological processes and the preservation of genetic diversity. In Africa the contribution of trees to total energy use is as high as 58%. The effect of such intense demand is to denude the land of wood over wide areas. Around one fishing center in the Sahel region, where the drying of 40,000 tons of fish consumes 130,000 tons of wood every year, deforestation extends as far as 100 kilometers. Fuel wood is now so scarce in Gambia that gathering it takes 360 woman-days a year per family (IUCN, 1980).

In Canada's High Arctic new requests for gas and oil exploration in the Lancaster Sound region have stimulated an exciting approach to the issue of resource use and conservation.

Ecologically the Sound is possibly the richest, most productive area in all the Arctic. Certainly, the long-term health of this special, indeed unique, environment is an important concern to us all. The Sound is also the entrance to the famous Northwest Passage, and as such, is a potential transit route for increased shipping if industrial development of any kind proceeds anywhere in the Arctic. Since the Sound also holds a hydro-carbon potential, we have important decisions to make about the safety of possible oil and gas exploration and development here as we pursue our national objective of energy self-sufficiency. And weighing heavily in the balance of our considerations for future uses of Lancaster Sound are the interests of the native Inuit who continue to depend on the area's resources (Government of Canada, 1980b).

The goals of the World Conservative Strategy can be seen as elements in two frameworks for assessing the environmental planning process: the concept of appropriateness and various approaches to natural resource management.

Environmental Planning Practice

APPROPRIATENESS. A draft green paper written to elicit public input to the planning process asked a series of questions designed to structure debate. Each of these "questions" includes aspects of issues that affect national interest, protection of the environment, life-style flexibility, and use of appropriate technology. These four themes are obviously interrelated, and one of the objectives of the environmental planning process is to organize and present "appropriate" information to the public and to decision makers so project proposals and alternatives can be properly designed, assessed, and monitored.

What is appropriate in one context, however, is not necessarily appropriate in another. Thus, *appropriate technology* is, by definition, appropriate to a given context. Consider the need for energy in two disparate contexts—the developing and developed regions of the world. In both contexts an "energy crisis" exists, although the nature of each crisis is profoundly different. In one, basic human needs for cooked food, warm shelter, and modest "mechanical advantage" derived from the industrial revolution are at issue; in the other, energy-intensive perceived needs for sophisticated transport, ubiquitous lighting, and sophisticated mechanical advantage are related to the "quality of life concept."

In the province of Szechwan, China, there are an estimated 5.5 million family biogas units. Construction costs for each are estimated to be $120,

which can be "paid back" in two years from the savings of domestic coal. This figure includes a price for labor, which is, nonetheless, highly participatory, as in the barn raisings of the North American West (Khosla, 1980).

Assuming that the average family unit in Szechwan served by family-sized biogas units is slightly larger than four, the equivalent of Canada's population is served by biogas at a total capital cost of $6.5 billion. This is very roughly one-third the cost of developing hydroelectric power in the James Bay region of Quebec, which is designed to serve, only partially, the energy-rich appetite of a population approximately one-quarter the size of that in Sichuan.

The James Bay project is worthy of note both because of its scale and the fact that it, like many other energy plants in Canada, is powered by a renewable resource, water. Notwithstanding the fact that the entire complex of energy projects planned by Hydro-Quebec for the James Bay region will be sustained on a renewable resource basis, the very scale of the projects and the nature of engineering works necessary to achieve them give rise to other issues of appropriateness.

The scale of the engineering works along the rivers that flow into James Bay is quite simply immense. To imagine the size of the La Grande 2 Dam, consider the following comparative facts (Fontaine, n.d.):

If the rock and gravel used to construct the main dam were placed in normal trucks, they would require 2 million trucks, each with an 11-cubic-meter capacity. If these were spaced every 33 meters, or 100 feet, the convoy would circle the equator more than one and a half times.

If the quantity of rock and gravel used to construct the dam and other retaining walls for the reservoir were spread on a highway from Portland, Maine, to Portland, Oregon, the entire highway would be raised by 1 meter.

The reservoir that results from the engineering design for this site alone is larger than the largest natural lake in Quebec. The entire reservoir surface created represents almost 60% of the surface of Lake Ontario, one of the five Great Lakes.

The debate over the social and natural impact of the James Bay project on Quebec as a whole has generated volumes of literature that are well beyond the scope of this presentation. I can, however, review the objectives of the Environment Department after five years of work to see to what extent they have been, or in fact can be, attained (SEBJ, 1980).

The first objective, "to evaluate through a recognized scientific approach the physical, chemical, and biological changes that would occur to the reservoir system," has to a certain degree been achieved. Major reservoirs have been filled and at least two major rivers have been diverted to augment the capacity of these reservoirs. Comparative figures before, during, and after this operation are available and have been analyzed. For the most part, however, these analyses are sectorial and do not yet provide a dynamic understanding of changes in terms of an interactive or systematic model of the environment.

As a consequence, the second objective, "to use this information as a means of managing the reservoirs and of rationalizing corrective measures," is only partially attainable. Insofar as the James Bay project exceeds our experiential knowledge, it, like the Arctic pilot project, must be considered a full-scale experiment. Management techniques and mitigating measures derived from sectorial analyses will most certainly be better informed than before, but they must also be considered experiments based on "best professional and scientific judgement."

The third objective, "to profit from this experience such that the methodology used to predict environmental impacts of future projects might be refined," is largely illusory at this time and for some time to come. In the absence of integrated experiential knowledge as expressed by a tested and verified model of the environmental milieu, the prospect of establishing a viable predictive capacity is probably marginal.

The issue of appropriateness as a criterion of environmental planning has been raised, but how does it apply to James Bay? Clearly the project has been based on the maintenance, if not the sustenance, of the postindustrial model of economic growth and development, consistent with the prevailing ideology of North America and of the developed world. Within this model every attempt has been made, in good faith, to monitor, evaluate, and mitigate environmental impacts that result from project implementation. At issue is the bald fact that we are engaged in a project whose very scale is beyond our acquired knowledge base and whose impacts cannot possibly be known within the relatively short time frame of project implementation.

APPROACHES TO NATURAL RESOURCE MANAGEMENT. The basic approach to natural resource management that underlies the World Conservation Strategy focuses on the concept that resources can and must be managed on the basis of sustainable yield. The premise of this concept is that each unit of land or water has a *carrying capacity*, that is, that there is a finite and determinable yield or productivity that can be derived from each unit; further use will seriously affect these units' continued ability to produce at the same or similar rates.

The concepts of *sustainable yield* and *carrying capacity*, however, have not received universal support from politicians, scientists, or planners. It is exceptionally difficult to establish rigorous values or quantities that can fairly reflect the productive capacity of a unit of land or water without resorting to "scientific perception" and/or "best professional judgment" by planning officials.

In an alternative to the carrying capacity model, one author proposes the "dead is dead" strategy, which contends that "once a threshold is violated, it is better to continue to concentrate activity in the "dead" area, than to encourage development on other areas that have not been forced across similar thresholds" (Mar, 1980).

Others have argued that "the overall volume of non-renewable resource natural wealth is constantly diminishing. However, this does not mean that the possibilities of meeting one or another of man's needs related to them are diminishing. Quite the contrary, as science progresses and production processes change, such possibilities increase, owing both to a more efficient use of each resource and exploitation of new resources of natural wealth and to the quest for fundamentally different solutions" (Federox and Novik, 1977).

Evidence in the World Conservation Strategy, however, does not support the premise that the possibilities of meeting one or another of man's needs are increasing, certainly not in the absence of concerted planning for sustainable use of renewable resources. Insofar as pressures on the use of the renewable resource base grow in direct relationship to increasing population and urbanization throughout the world, much more attention must be paid to methods that will assure achievement of the World Conservation Strategy's goals. The environmental planning process is one. It is an essential tool for prediction, design, management, and monitoring of the changes to the landscape that result from development activity. Just as value systems change from culture to culture and the biogeographic context varies across longitudes and latitudes, so does the essential information that animates the environmental planning process. The objectives of this process, however, are to organize the ways and means of sustaining development. The means may vary to suit particular contexts, but the objective remains constant, as does our need to support the goals of the World Conservation Strategy, upon which the environmental planning process is based.

Discussion

QUESTION. Even five years ago people would have been talking about plans with a capital *P*. I find it very interesting that nobody has talked about a plan, particularly a master plan—the most villainous of all plans—something that runs to 500 pages with three appendices to which nobody pays attention. Instead you have been talking about tactics, suggesting that there are a whole range of them. Is this a new thought in planning?

ANSWER. Many people have suggested that master plans are limiting, maybe even constraining. In some cases they are obviously nonsensical. A series of colored diagrams with lines on them is only one of many planning tactics, and in many cases it can be a useful one. But even if such plans are necessary, they are certainly not sufficient.

There have also been disasters with legislation that stipulated, "There shall be a plan, and the plan shall be implemented." There is so much inflexibility in such processes that when something changes in a way they did not anticipate, the plans are abandoned. They no longer represent or mirror policy. So we tend not to talk about "the plan" anymore, we talk about "the policy."

QUESTION. Has there ever been one case where people have gone back and reviewed an environmental evaluation? I have a strong suspicion that this new tool may not be terribly more effective than the old tool, but that we do not know it yet because nobody has gone out and asked whether we really did predict the significant impacts. I wonder if anyone has studied the process to see whether people choose the right impacts to evaluate and whether they predict the right levels of the impacts they do measure.

ANSWER. I am as scared of environmental impact statements (EISs) as I am of master plans. There is a ritual about naming things. When you have named something, you have captured it; you have boxed it and packaged it; you believe you have got it. Environmental impact statements have credibility these days; they are the going entity. There is a lot of thought being put into designing them, generating them, assessing them. I suspect that one of the legacies of the environmental impact assessment era will be that we will have generated more scientific data about more places than we will ever be able to use.

Practically, however, a postmortem on an EIS is a contradiction in terms. As things are now, an EIS is seen as a green light; once it has been accepted you have permission to do what you are proposing to do. Because it has been evaluated as okay, the process says it must be okay, and there is no need to review it. It is a beautifully tautological argument— that is its strength, it is tautological. The concept that in the very doing of something things can, and inevitably do, change is not integrated into the EIS process. It is not yet seen as an evaluation and a monitoring process.

Even the monitoring process is a very dangerous issue. In James Bay they have monitored, and it has cost phenomenal amounts of money. They have put money into fish nets to catch fish running up and down the river, to measure turbidity, acidity, salinity. The dimensions are absolutely staggering! But nobody has put it all together. What you have is an enormous pile of data.

The process will have to become a true process. Right now we have a product-by-product, step-by-step approach. It is static in time. You cannot really evaluate whether you have second-guessed all the impacts because things are *never* built as proposed. We need a process that monitors a project continually from start to finish.

QUESTION. I am very frustrated by the planning process. It leaves no room to question basic values. I am very interested in the examples you gave. In most cases energy self-sufficiency and energy generation are viewed as higher national priorities than the protection of marine mammals and other living resources. It seems you can do all the planning you want to, but the policy bias remains. How can you elevate the importance of environmental elements in planning situations?

ANSWER. The Arctic pilot project that I mentioned sent shivers through the system. The government reacted by saying we will do nothing until we have all parties and issues accounted for. A planning process was instituted that gathered data from both conservation and development sectors, and from a whole range of others. It was an intergovernmental process whose initial product was called a "draft green paper." It said there are four possible scenarios:

1. Drill first, and after the drillers get finished staking their territories, leave the rest as parks
2. Get the parks first, drill second
3. Don't drill
4. Don't have any parks.

In each of those scenarios, there are three issues at stake: energy self-sufficiency, nature conservation, and socioeconomic factors relating to the life-style of native people, the Inuit.

A national workshop will be held at which these four scenarios will simply be put on the table. People will be invited to say what they think the best strategy option is and why. These comments will then be collated and deposited on the minister's desk, where a political decision on the scenario will be made. This process might not solve the problem, but at least the Canadian government is trying to find out what the public value judgments are. It is trying to structure this debate to avoid polar confrontation if possible.

Still, I am nervous about the process, because I think we are dealing with too many things that we truly do not understand. We do not have any comparative experiential base against which we can judge the impacts. There is also a snowball effect. Once the first drill goes in, or the first park, you have changed the entire problem. If the park goes in first, there will be a lot of ammunition against drilling, and vice versa. This is part of the whole evolving perception of the environmental planning process: Things do not stay stable, they are always changing.

QUESTION. I always thought of conservation as a decentralized process. What do you think of decentralized versus centralized conservation? Which direction do you think things should go?

ANSWER. There is one obvious point that people have not thought about as much as they should. You cannot be involved in environmental planning unless you have a local base of support. You might be able to think about the problems in Thailand abstractly, but I believe you cannot begin to deal with those problems until you have had some local experience, commitment, and involvement.

I have worked in rural Quebec, where the municipalities do not have enough professional expertise to comply with the laws that require them to generate plans. A very exciting thing happened to me in one of these towns. They had to generate a development strategy to meet a deadline about three weeks away when they contacted me. We got the best hunter in the municipality to tell us where all the wildlife was. It took him exactly three hours of work at a kitchen table. He knew where the deer were, where their winter and summer habitats were, where the links between the two were. He drew a map that eighteen professional ecologists working for $1,000 a day for six months could never have delivered. That hunter was also one of the better conservationists I have known. He did not want to ruin his livelihood. Similarly, we called on an elderly woman who had lived in the municipality all her life and knew all the local history.

There are an enormous number of people with expertise in every community. There are always ways of getting that knowledge out on the table and using it to create conservation and development strategies. I cannot imagine being involved with conservation without being involved locally. No one should work on a national or international scale without local experience. The linkage must be from the bottom up.

QUESTION. It seems to me that the size of the impact grows geometrically as the size of a project grows arithmetically. It also seems that the expertise of the planner working on large projects has to be enormous. What do you think about a project with as much negative impact as the Aswan Dam?

ANSWER. Many of the impacts at Aswan were known and the situation was clearly identified as disastrous beforehand. We have to get back to the question of values. A dam is a tangible solution to a problem. You can build it, and you can say that you have solved the problem. The productivity of a dam is calculated on a fifty-year basis. In many situations the management of the watershed flowing into that dam is not considered a relevant part of the problem. If watershed forests are cut for their immediate timber value, siltation occurs that can cut the life span of the dam from fifty to ten years. Your original equations lead to financial disaster under these circumstances. You have economic factors and you have perceptual factors. The fact is, people perceive the dam to be a positive solution mainly because it is big.

QUESTION. Will we continue using environmental impact statements?

ANSWER. It depends on whether you are looking for a procedure-oriented environmental planning process or a result-oriented one. The last decade has been incredibly successful at integrating environmental quality concerns into the development process. Substantively, it may not have changed the ultimate decision making to the satisfaction of many, but now conservation is an integrated part of the review. Ten years ago environmental assessment was not even a consideration in project planning.

One of the problems is that we still formulate the wrong proposals to begin with. The environmental impact process is not enough. We should start doing an EIS when we have a goal and want to formulate alternative approaches, not once we already have a specific proposal.

QUESTION. How did the World Conservation Strategy planning section end up the way it is? To me it looks very similar to the kind of system you say you have doubts about. It seems the public comes into it down the way, not at the beginning.

ANSWER. The fact that participation exists at all is a very serious political risk for a world conservation unit like IUCN. Do not believe for a moment that every government in the world is dying to get its population to participate in these decisions. I consider the planning section an accurate statement of the way we were thinking a little while ago. I do not think we can ever get the complete process as I have described it. However, I am ridiculously optimistic that if we can buy a little bit of time, we can show some real examples of a worthwhile process. I did not talk much about another conception of environmental planning, ecodevelopment, because Raymond Dasmann explains it well (see Chapter 2). There are going to be positive results from this approach. Groups are now doing their own small-scale planning. This process will have to be evaluated in terms of quality of life rather than gross national product, but for vast numbers of people it will mean better lives. That is the shift I am looking for.

Building Support
for Environmental
Education

WILLIAM STAPP

I want to start with a quotation from an astronaut. It could have been made by a person from any country, but this happened to be an American astronaut, Russell Schweickart. While circling the earth in a spaceship, he wrote the following:

Up there you go around each hour and a half,
Time after time—you wake up over the Mid-east, over North America.
As you eat breakfast you look out—and there is the Mediterranean Area—
And you go down across Africa and out over the Indian Ocean and look
up at that great sub-continent of India—and then you look for those
friendly things: Los Angeles and Phoenix.

You can't imagine how many borders and boundaries you crossed—
and you don't even see them—you know there are millions of people
quarreling with each other over some imaginary line that you can't see.
From where you see it, the thing is whole, and it is so beautiful.
And you wish you could take one from each side in hand and say,
"Look at it from this perspective. Look at that. Let's share our
ideas and information."

*The bright blue and white Christmas tree ornament in that
    black sky,
that infinite universe, really comes through—it becomes so
    small and
so fragile, in such a precious little spot in the universe, that
    you
can block it out with your thumb and you realize that on
    that small spot,
that little blue and white thing, is everything that means
    everything to
you. All of history and music and poetry and art and war
    and death and
birth and love, tears, natural and human resources, all of
    it is in that
little spot out there that you can cover with your thumb.*

Participation and Education

Environmental problems exist in all countries of
the world and at every state of economic development and political ideology. Developing countries
frequently experience problems associated with underdevelopment or poorly planned development—
farming techniques leading to soil erosion and depletion, improper management of forest resources,
poor health and nutrition, vulnerability to natural
disasters, and the lack of educational programs to
help resolve these problems. Some developing
countries have adopted inappropriate measures
based on short-term gains or unsuited to existing
situations. These strategies have led to the rapid
depletion of resources, increased pollution, and, in
some instances, the spread of disease. Many developed countries also are faced with severe environmental problems—industrial pollution, overexploitation of resources, and the variety of social and
physical problems confronting metropolitan areas.

When development programs are not planned
adequately, they result in the deterioration of natural and human resources. The litany includes the
reduction in quantity and quality of mineral resources; the destruction of ecologically important
lands, forests, and aquatic sites; biological pollution
by organisms that cause disease in humans; chemical contamination from effluents, pesticides, or
other materials; and physical pollution, such as
noise, silting, thermal wastes, or visual blight.

Such environmental problems cannot be resolved by looking at only technological solutions.

Consideration must also be given to the social and
economic factors that led to them. We will be facing the same environmental problems in the future
and breeding new ones, until we identify their
causes and develop programs designed to help resolve them. It is evident that there can be no hope
of finding workable solutions to environmental
problems until education is modified to enable people from all walks of life to comprehend the fundamental interaction between humans and their environment.

Within the attitudes of our population lie the behavioral roots of such problems as pollution, wasted
energy, and destruction of the environment. We
lack a global ethic encompassing the world environment and espousing attitudes and behaviors of
individuals and societies consonant with humanity's role in the biosphere. I do not think of environmental education in terms of a product. The process is what is important, getting people to
understand the implications of their various possibilities for action.

What are the starting points for environmental
education? One is individual behavior. It is important to think in terms of individual people, the
choices they make, and their initial perceptions of
the natural environment.

It is also important to remember the balance between environmental and nonenvironmental
needs. From an ecological and economic perspective, many externalities can be involved in creating
the products society demands. Consider the problem of air pollution. Many pollutants fall out of the
air very quickly. Others carry many miles, and their
impacts have wide international implications.
Often the people most affected by industrial pollution are the politically and economically disenfran-
chised who live nearby. They are paying an extra
price for our consumer goods.

Another fundamental concern is the timing of
educational programs. This is a function of people's
present understanding. We must look at the traditional life-style of the audience for a program. By
asking them the right questions, at the right time,
we can guide them toward an understanding of
their present relationship with the environment.

The third starting point is the "values problem."
Education is never value free. We must examine

FIG. 8.1. *Formal schooling is only one aspect of education. A village planning meeting in Botswana engages local participation in environmental decisions. Photograph by M.J. Odell.*

beliefs and attitudes as well as behavior. Values have two components: cognitive and emotional. Those who have strong feelings and strong knowledge are equipped to act. People's present values are a function of their culture and experience. Often the most direct sort of environmental education is to present learners with a new experience. In San Francisco a coercive sort of behavior modification was used to get people to start car-pooling. I have trouble with this approach, because it can create a negative attitude toward the new experience. It forces people to act before they are ready and can keep them from discovering whether they really value the new idea.

There are other ways to change values. You can take a "moral" approach: Someone decides what is right for the whole society and begins to preach it. I am not an advocate of this approach either when teaching adults. Because they do not have the ability to reason yet, there is a place for moralizing with very young children. We tell them they must not fight with brothers and sisters, for example. But even at those ages modeling is important. Teachers and parents must demonstrate the proper behavior resulting from a set of values. This is an important part of teaching, whatever the student's age.

One of the most recent approaches to environmental education to gain attention is *value stratification*. It is the attempt to look at issues with an eye toward alternative solutions. If congestion in city streets is a problem, what are the alternative solutions? People can walk, bicycle, car-pool, or use mass transit. What are the implications of these alternatives? Or to put it another way, how do the "students" feel about them? If they are not making a free choice to adapt to a new behavior, their feelings will be colored. The teacher's first job is to provide a number of experiences in an environment where the students can be receptive to them. The second part of this process is to get the students to think in terms of the implications of their choices and their feelings about them. The teacher's job is not to say what is right or wrong, but to help the students think through the issues for themselves.

It is also necessary to examine the *logical problem-solving approach*. To me, the essence of problem solving is helping people recognize their problems. This ability is itself a major skill. The next step is definition, asking What is the real issue? The problem must be defined clearly, succinctly, accurately. What *is* the energy issue? What *is* the food issue? After that comes information—discovering ecological, economic, political, and social data that are necessary to come to terms with the problem. Next come analysis and the formulation of alternative solutions. At this point one can begin to question the implications of those alternatives. But once you begin stating alternatives and examining their implications, you are back to examining values. Eventually this whole process leads to a plan. As the plan unfolds and is implemented you have to test your expectations against what really happens.

During a problem-solving process, you have values and knowledge coming in, and you begin to make use of individual skills, which should lead to better citizen participation. For learning to be successful, the problems you deal with must be relevant to the learners and within their range of comprehension. If you select problems in the range of the learners, you begin to help them learn. You become a teacher, a facilitator. The teacher need not lecture, but he must provide a proper learning environment. As their skills become developed and they grow in self-concept, students are more inclined to continue work on these problems by themselves. Many people in inner-city environments do not feel they can bring about change, often because they have not been successful in the past. A good teacher will help them evaluate their past experiences and develop new strategies.

The fundamental goal of environmental education is to equip people to think in terms of the future. Maybe you start talking about your own neighborhood, then pretty soon the discussion broadens to include all of your region, soon after that the topic becomes the future of your country, and then the whole world. Once you have come to terms with questions of freedom, awareness, and understanding, it is possible to focus on future events and the relationships among them. Those events may have implications not only locally, but also regionally and even globally. But the further removed in time and distance the implications are, the harder it is to stretch people's awareness to encompass them.

The World Conservation Strategy recognized that ultimately the behavior of entire societies toward the biosphere must be transformed if achievement of conservation objectives is to be assured. Let's look at how the strategy attempts this.

The UNESCO-UNEP Program

The concept of environmental education put forth by the World Conservation Strategy is a continuation of the Environmental Education Program, which was formulated by the U.N. Educational, Scientific, and Cultural Organization (UNESCO) and the U.N. Environment Program (UNEP) in 1974. This program was a direct result of a recommendation at the U.N. Conference on the Human Environment in Stockholm in 1972, which stated that "organizations of the United Nations system, especially UNESCO, should establish an international program in environmental education, interdisciplinary in approach, in-school and out-of-school, encompassing all levels of education and directed toward the general public, in particular the ordinary citizen living in rural and urban areas, youth and adult alike, with a view to educating people as to simple steps one might take to manage and control one's environment." Furthermore, the conference called for UNESCO to work with all appropriate U.N. agencies, international nongovernmental organizations, and the 148 member nations to develop a framework for furthering environmental education internationally. A direct result of this recommendation was an intergovernmental conference on environmental education held in Tbilisi, USSR, in October 1977 to formulate and adopt international, regional, and national policies on environmental education.

In October 1974 UNESCO convened a consultation meeting. Representatives from many U.N. agencies, international governmental environmental education organizations, and experts in environmental education from each region of the world were present. An outcome of this meeting was a UNESCO proposal, later approved, to UNEP for $2 million over three years. The resulting UNESCO-UNEP effort was designed to help develop an international program in environmental education; promote the international exchange of information on this subject; coordinate research in teaching and learning; and formulate and assess new methods, materials, and programs.

New Directions

The World Conservation Strategy builds on the foundation of the UNESCO-UNEP Environmental Education Program and elaborates on a point not strongly emphasized in previous movements—public participation in conservation and development decisions. Public participation is particularly important as a way to ensure that decisions reflect the experiences and wishes of the people affected by them. Furthermore, the strategy recognizes other benefits of public participation:

1. Plans are more sensitive to local needs
2. Local people are more apt to contribute time, materials, and labor to carry out plans
3. Implementation is smoother and quicker once understanding and assent have been generated
4. Integration of activities and services is more effective and complete
5. Political support is greater if some of the ideas are generated by local people.

The strategy also recognizes that school curricula should include environmental education as an intrinsic part of other subjects, and as a separate subject, and that inexpensive teaching materials should be prepared for instructors. It also encourages greater use of mass media, national parks, museums, and other learning environments that can enrich instructional programs.

A Case Study:
Schistosomiasis in Egypt

I would like to conclude with a case study to emphasize the importance of understanding the cultural dimensions of effective education programs. I will focus on schistosomiasis in Egypt.

The disease schistosomiasis is widespread in Africa. In Egypt it is considered the number one public health problem, and in Nigeria it is second only to malaria. Patients do not usually die from schistosomiasis. In most countries the death rate directly attributed to this disease is less than 3%. But infected individuals tend to become weak and emaciated in the later stages of their illness, and they may succumb to other diseases. In areas where it is severe, schistosomiasis drains the energy of the people. It has been estimated that Egypt has suffered a 30% loss in economic productivity as a result of this one disease. One parasitologist feels that Egypt will never be a highly productive nation until the disease is brought under control.

Schistosomiasis is caused by a blood fluke. The adult fluke lives chiefly in the blood vessels of the bladder and intestine. The eggs work their way through the vessels into the bladder and intestine and are eliminated in the urine or feces. If this waste material was removed by an adequate sewage system or deposited in a dry area, the eggs would not hatch and the schistosomes' life cycle could not be completed. But if the eggs reach fresh water they hatch into ciliated larvae, called *miracidia*.

These larvae must encounter the right kind of snail within approximately twenty-four hours to continue their development. If the proper host is found, the miracidium penetrates the snail's body and feeds on its tissues. Within the snail host the ciliated larva is transformed into a *sporocyst*. This saclike structure produces internal buds called *cercariae*, which are released in about four to six weeks. These fork-tailed larvae make their way out of the snail's body and swim in the water. If people drink water containing cercariae or if these larvae penetrate the skin while people are bathing or

FIG. 8.2. *A community water tank in Egypt: one step toward breaking the life cycle of the parasite that causes schistosomiasis. Photograph by M.J. Odell.*

working in infected water, the cercariae enter their blood vessels and are transported to the vessels of the intestine or urinary bladder, where they transform to mature blood flukes. The cycle from cercarium to adult takes six weeks. The life cycle of the parents continues when the females force eggs through the smaller blood vessels of the bladder or intestine. Many eggs fail to reach the outside; if they remain in tissues of the human body, they produce lesions and sores.

In the past schistosomiasis has taken its natural course and spread into areas where conditions for it to complete its life cycle were present. In recent years economic trends have played a major role in creating new snail habitats in areas that had been largely uninhabited. As a result of new irrigation, water storage dams, water slowed down by the Aswan Dam, fish farming developments, and general population movements, the disease has spread rapidly. Millions of new cases of schistosomiasis are the result.

Any program directed at educating people to control schistosomiasis must consider the cultural factors that lead to the disease as well as the technological forces that have spread it. The religion that has had the greatest influence in North Africa is Muhammadanism. The Islamic people pray five times a day and they wash before each prayer. In the rural areas prayers normally take place on straw-covered areas on the banks of canals and rivers. The practice of performing ablutions after both defecation and urination and before each prayer is just about perfect for the life cycle of the parasite that causes schistosomiasis.

Sheikhs have attempted to encourage the Islamic people to carry small bottles of well water to a dry site to perform their ablutions after excretion and before prayers, but the results have not been very successful. It is very difficult to break a tradition and habit that have been followed for many lifetimes. In one village in Egypt, well water was provided for ablutions, but the sheikh forbade the people to use it—he had not been consulted by the authorities who distributed the water, so he felt that his leadership was being usurped. The people obeyed their sheikh and returned to their normal use of canal water.

As one observes the daily habits of the Islamic people and then considers their numbers and the area they inhabit, one begins to realize that any program aimed at controlling schistosomiasis must consider both the Moslem religion and the local power structure. Programs fail when people are not made ready to receive them. Progress is being made in the urban areas of Africa to furnish information about prevention, detection, and treatment of schistosomiasis. Many schoolchildren are receiving instruction in their science courses, and adults are being reached through pamphlets, radio, newspaper articles, and films. However, in the rural areas there are not enough schools to handle all the children of school age. There are accounts of schools with capacities of 60 that turn away over 200 children each year. In many schools the major part of the day is spent reciting the Koran, and little or no time is given to health or other educational issues. Education is faced with the problem of breaking down the barrier between central authority and the village people and helping to develop within the people a willingness and desire to improve their sanitation habits. If this can be accomplished, then progress in controlling schistosomiasis in Africa will follow.

Although steady progress is being made in environmental education internationally, some major challenges still lie ahead (see Aldrich, 1977). Greater efforts must be made to coordinate and to assist agencies in the U.N. system and other international organizations in identifying ways to further such education. And very little action may occur unless national plans are developed to facilitate the establishment and operation of environmental education programs and activities.

It is through environmental education that a foundation for an environmentally literate citizenry can be laid. This foundation, and continued environmental education programs, should make it possible to develop new knowledge and skills, values and attitudes, in a drive toward a better environment and, indeed, toward a higher quality of life for present and future generations.

Discussion

QUESTION. It seems as though there was much more emphasis on environmental education a while ago than there is now. Are we losing ground?

ANSWER. There was a big surge of emphasis on environmental education, particularly in the United States, around 1969. At that time environmental affairs were receiving much more coverage in the media and there was a correspondingly greater demand for more formal environmental education. As a result, the U.S. government started a program to develop statewide environmental education plans in 1971. There are now forty-three states with such plans. These were voluntary programs until the National Environment Education Act made the plans mandatory. The act lapsed, however, and so did the programs.

Many countries still make environmental education a high priority. For example, the government of Peru has established an exceptionally fine plan. There are also very interesting efforts in other parts of Latin America, Africa, and Asia. Eastern European countries are doing a fabulous job. Recent meetings in Hungary brought all the eastern European countries together to discuss the problem. The Soviets are taking an especially strong interest in environmental education. These countries have youth groups monitoring the environment in addition to the official government programs. All their industries have to respond to these young people's questions about their practices. Many good and innovative things are occurring in other parts of the world.

It is quite possible to see the United States and western Europe as the new backwaters of environmental awareness. Other regions are moving ahead quickly. The World Conservation Strategy is taking hold in developing countries, particularly in nations such as China and Indonesia. They are quite clearly linking conservation education with development.

QUESTION. Most of your presentation has been about education in the traditional sense, but you have at least hinted at environmental education as a component of projects with more direct physical change as their main goal. Could you expand on this idea?

ANSWER. The environmental impact report, for example, is a very fine educational tool. By going through the process of producing the report, the developer learns about what will happen at the site as the project is carried out. When the law requiring impact statements first passed, I was presented with a thick environmental impact statement on the building of a large subway along a major highway in our town. In looking through it, I found the authors really had not thought much about the environment. They either ignored problems deliberately, or they did not have enough information on them. The wells for the town's entire water supply were within 200 feet of the proposed station, but they had not even been mentioned. However, I have noticed a tremendous improvement in the quality of impact statements in recent years. I think people in development have become educated because we insisted on passing legislation to make sure that they stopped and thought before they began a project.

Citizen participation has been an important part of this process: I walked down those tracks with the engineers and scientists doing the impact statement. They went back and put those wells on their project maps. There is room for the citizen to participate in the education of the professionals who are making development decisions. Citizen action is an extremely important part of environmental education.

I am very concerned that we are not reaching children through our school systems. We are not making them think about water supplies or open space needs. So it is no surprise when technical documents do not mention these aspects of the environment. Today we have to teach environmental education and awareness to people who are already developers, industrialists, and economists. It would be better to begin the teaching much earlier.

QUESTION. In New England we have been trying to find a site for a coastal oil refinery for years. It seems clear to me that a number of people would not accept any site, even the best. I wonder whether some environmentalists are really interested in making good choices about development or whether they become involved in issues such as this one to obstruct the process of making good decisions. Could you react to this?

ANSWER. The basic issue is, What is a good choice? From your perspective, a good choice may be based on utilitarian considerations; from someone else's, a good choice may reflect a more holistic or global ethic. If you are willing to consider a global ethic, the actions you choose are likely to reflect that framework. It may be that you cannot build an oil refinery on the New England coast, for example. The possible effects on the fishing industry may be more than you want to risk. A major oil spill on Georges Bank could upset the balance of a global food resource. I am not sure that there are "good" choices except in a very subjective sense. Environmental education must show students that people's final actions are always at least partially a result of their environmental ethic, be it utilitarian, scientific, or aesthetic. People must understand that actions are based on their fundamental values.

QUESTION. Each education project should have an evaluation as part of the process. You have to measure yourself against your objectives. Does IUCN use standard criteria in their evaluations?

ANSWER. UNESCO has released a guide to evaluating educational projects. It is not a single model to use, but rather a list of a variety of factors that must be considered. Its most important emphasis is that the people who develop programs and the people served by these programs must have the most important voices in any evaluation process. In the past there was very little project evaluation, especially in the Third World. Then we went through a stage where evaluation was always tacked on at the very end. Now we have begun helping people examine their objectives and strategies from the start of a project.

Conservation-Based Rural Development

DAVID WESTERN

In looking through the World Conservation Strategy one cannot fail to be concerned by how little space is devoted to rural development and by how familiar its "conservation-based" solutions appear to be. Although the problems arising from rapid and sustained human growth in rural areas are elaborated in depressing detail, the actions recommended to forestall the seemingly inevitable loss of most remaining biological resources are given little attention. The solutions offered—such as livestock reductions, increased efficiency of food production, irrigation, provision of alternative areas for human settlement, and so forth—suggest little that has not long been tried by development agencies.

Why does the single most important challenge confronting conservation—the fate of biological resources within the rural areas, which implies most of earth's land surface—receive so little attention and elicit so few new ideas? Perhaps because of our despair at the rapid expansion of human populations and our pessimism about the available solutions; certainly it is not surprising that conservationists, who are usually trained in biological sciences, are frustrated because simple ecological principles are either not appreciated or are inadequately applied (Ehrlich and Ehrlich, 1970; Dasmann, Milton, and Freeman, 1973). And certainly it is rash to argue otherwise; too many planning disasters testify to the contrary. But it is equally myopic for conservationists to deny that most solutions they offer are already standard fare among, for instance, agriculturalists, for whom soil conservation, livestock reduction on pastoral lands, irrigation, the relocation of settlements away from productive croplands, and efforts to increase food production are all fundamental tools in the devel-

FIG. 9.1. *Hell's Gate Gorge in Kenya. While they can be strikingly beautiful, a few isolated parks and reserves will not maintain existing biological diversity. Photograph by C. W. Leahy.*

opment kit and have all long been applied and frequently failed. We can almost caricature the conservationists pointing out classic failures and how much remains to be done, while in contradiction the agronomists point out their great successes to date. Ironically, both can argue the record in their favor. In parts of Southeast Asia agricultural innovations have nudged the rate of crop production ahead of population increase, a trend that is likely to continue; in Africa, on the other hand, population growth will pull ahead of food production for the foreseeable future (Barr, 1981).

For the most part, conservationists and rural development specialists have historically had separate arenas of activity, the one concentrating on lands beyond the development fringe reserved for conservation and recreational use, the other endeavoring to increase human welfare within both urban and rural areas. Overlap in their activities has been confined to state or federal lands, especially forest reserves, where principles of sustained resource use have long been applied to a greater or lesser degree, and where wilderness recreation has, as in the

United States, supplemented the limited land available within national parks for preservation and recreational use (Nash, 1967; Runte, 1979). The area of overlap is, I consider, now expanding with the realization among conservationists that a few isolated national parks and forest reserves will be inadequate to maintain earth's existing biological diversity.

Integrated land management in the rural areas is undeniably the best hope we have to retain the majority of species in abundance and at affordable costs. If we fail to institute conservation principles and practices there, humanity will compress so tightly around most parks and reserves that few will survive the suffocation in any form that we presently call natural.

For this reason I regard developments in rural areas as the preeminent conservation challenge in the years ahead, and this is why I believe conservationists cannot afford to stand apart from the human predicament, expounding principles of development without asking how they might practically be accomplished to demonstrable human benefit. To achieve those ends, the conservation minded must become more conversant with rural development, more conscious of the human issues involved, and more aware of how they influence the outcome of conservation-based programs.

In this chapter I will therefore concentrate on the problems of rural development to underscore the nature of the constraints involved, in the belief that we lack less ideas about conservation-based development than understanding of how to incorporate them into the everyday plans of particular societies on a time scale commensurate with their expectations of some return. I will then highlight a few techniques that offer prospects of improving the imbalance of resource use, and, finally, I will give a few examples of solutions that derive from a conservation approach.

Developmental Problems

Most ecologists tend to perceive the issue of development, especially in the less developed countries (LDCs), as one of absolute population growth. Demographers and development specialists perhaps more realistically conceive the problem as one of relative growth, that is, the rate of increase compared with the rate of resource production. Historical patterns of growth in Europe and North America demonstrate that, provided production exceeds population growth, societies can become more affluent, work shorter hours, and have more leisure time available to them (Sauvey, 1966)—a combination of factors that leads to increased recreation, especially outdoor pastimes (Brightbill, 1960). Indeed, it has been argued that the conservation movement in North America was an outcome of population growth, the pressures of urban living, and the realization that outdoor recreational opportunities were fast disappearing with the wilderness (Nash, 1967).

The historical patterns in the north temperate countries are likely to be reflected to some degree in the LDCs; with increasing per capita wealth, urbanization, leisure time, transportation, and separation from nature there is likely to be a greater demand for and fascination with wilderness areas. Paradoxically, conservation of nature for its own sake, or for the recreational opportunities it offers, is an outgrowth of urban and rural development. Even within urban and rural living space affluence fosters a need for more space, more aesthetic surroundings, and less pollution. In Third World countries the great preponderance of urban, compared with rural, citizens who visit parks and reserves (even though the overall use is yet small) suggests a similar emerging trend. However, any hope that the pattern of the developed world can be readily repeated on a global scale is bedeviled by new forces and constraints, which, unless solved, will inevitably destroy many of our remaining natural resources. We can gain some perspective on the issues by considering past and present growth rates.

During the phase of maximum population expansion in the northern countries, annual growth rates did not exceed 1.3% (Coale, 1974); in contrast, agricultural and industrial growth have exceeded that level appreciably ever since the industrial revolution, leading to a rapid increase in wealth. Growth in the LDCs now averages 2.5% annually, and in a number of countries it is 4% (Coale, 1974), whereas food production has only grown 2.4% on the average over the past three decades (Barr, 1981). Over that same period per capita food production in the developed world increased at between two and three times the rate it did in the LDCs.

Why should there be such a large difference between the developed and newly developing worlds in the relative growth of populations and resources? Clearly, the answer is crucial to any attempts to bring production up or growth down to a point where per capita wealth can increase sufficiently fast to cause a demographic transition from large subsistence families to smaller and wealthier ones. The longer such a transition takes, the larger the ultimate population on earth and the fewer the resources per individual.

The problems of replicating the historic demographic transition in the LDCs are varied and not easily remedied. Population expansion in Europe found at least minor relief in emigration to the

Americas, Australasia, and Africa, while resource depletion was offset by imports from these yet unexploited continents. Industrial expansion, fueled initially by the vast organic reserves of the temperate forests, accelerated with the use of coal and finally oil. Now, as the emergent nations enter the industrial marketplace, over 75% of all commercial energy is consumed by the developed countries, although they make up less than 25% of the world's population (Leach, 1979). Further, diminishing fuel reserves and escalating demand over the last decade have driven oil prices upward by an order of magnitude. And, because more than 80% of the world's coal supplies lie beneath the soils of the northern nations (Sassin, 1980), the problem will be further compounded for the LDCs once oil supplies dry up and coal once again becomes the major energy source. Energy intensification has been the conventional pathway to greater agricultural productivity, through sources required for both mechanization and fertilizers (Brown, 1981). Thus, crop yields may average some five times more in the developed than developing countries, but consume nearly fifteen times as much energy (Pimentel and Pimentel, 1979). So, unless new technologies are found to break the food-energy equation, it is difficult to imagine how the LDCs can achieve greater productivity on suitable land and halt the further depletion of biological resources.

The unprecedented population growth of newly developing countries also arises from novel circumstances. Whereas in the past death rates remained relatively high despite improvements in living conditions, modern drugs are available to emergent nations and have substantially improved infant survivorship well in advance of a decline in birthrates responding to changing living standards and expectations (Coale, 1974). In subsistence societies the need for labor to produce food constrains a rapid downturn in the birthrate (Ominde, 1972). Equally daunting is the thought that even if birthrates plunged immediately to replacement levels, the population of most LDCs would still double by the turn of the century as an inevitable outcome of the number of children yet to begin having children of their own.

Among the additional catch-up problems can be included those of new nations having to break into well-established trading networks (Leontief, 1980); of a poor infrastructure precluding rapid internal advances in marketing; of transport and costs of acquiring fertilizers, insecticides, simple mechanical equipment; and so forth. And finally, within the poorer nations investments are attracted to regions of greatest economic return, typically the high-potential regions, where birthrates have in many cases already begun to fall. The more expansive and populous regions are thus given less attention; yet they are the areas of greatest population growth, resource needs, and expansion of poor agricultural practices.

Environmental Impact

The environmental impact of expanding human numbers and activities is likely to destroy a considerable portion of the world's natural resources before the slackening tide of numbers and improved production methods stem the loss; indeed, the irony is that the short-term need to produce an adequate living for today's population by using extensive methods of agricultural production is at the expense of the long-term global carrying capacity. There are several interrelated reasons for this. First, most of the population increase is now located in the developing world where, over the next decade, 65 million people will be added to the 2.3 billion already hard-pressed for a living; in contrast, only 6 million will be added to the 790 million enjoying a higher standard of living in the developed countries (United Nations, 1980). Inevitably, then, most environmental pressure will occur where agricultural productivity is low, the means to intensify are limited, and a farmer's investment horizons are restricted by the need to feed a family already at the limits of subsistence.

Immediacy of human needs, limited investment options, and the delays involved in reaping the greater gains of investment impose far greater impact on land and biological resources today than formerly, when space permitted migration in the face of diminishing productivity and land degradation. In other words, current patterns of resource degradation are not new, but their intensity is historically unprecedented. It is worth considering briefly typical agricultural procedures, for these age-old patterns have a bearing on which principles of development can be applied.

Under traditional shifting agriculture, human impact on the landscape and ecosystem is quite evident, if limited. Land must be cleared, typically by burning, and the soils tilled to diminish plant competitors (Allen, 1965). Consequently, the farmers' short-term localized impact strips plant cover, eliminates animal competitors, and locally increases erosion rates. Both erosion and continual cropping soon deplete soil fertility (Nye and Greenland, 1965) except in areas with seasonal replenishment, such as floodplains. Many shifting cultivators move to new sites even before the depletion of soil fertility simply because it is cheaper to clear a new field by burning than to contain the reinvasion of weeds on an established site. During the course of occupation local wood stocks are cut down for firewood, building poles, and fences. Within foraging distance of the fields, wildlife are often extensively hunted to supplement the farmers' diet and seasonal food shortages.

Extensive pastoralism, which still occupies about a quarter of the tropical lands too dry for agriculture, is also traditionally a migratory life-style in which, once again, loss of plant cover and greener pastures elsewhere are the stimuli for movement (Western, 1982). Erosion and nutrient depletion resulting from the reduced ground cover are limited in intensity. And, as in shifting cultivation, wildlife is often a crucial supplement to the normal subsistence diet. Competition by domestic stock reduces biological diversity and wildlife numbers, but it is seldom sufficient to supplant them entirely.

The dominant forms of land use under traditional subsistence agriculture had, I believe, a similar environmental impact to those of today—soil erosion, nutrient depletion, reduction of woody cover, and the elimination of plant and wildlife competitors—which, in short, lead to a localized reduction in biological diversity. But from the traditional human perspective, biological diversity is of little significance unless it contributes to the welfare of subsistence cultivators and pastoralists, whose priority concerns are food and water. The survivorship of families who manage the land for personal wealth inevitably favors their expansion, even though biological diversity and gross produc-

tivity are reduced and seasonality and instability of the environment are increased.

The maintenance of high biological diversity is generally at the expense of human subsistence productivity, and its costs are therefore measured in human, rather than environmental, terms. We can perceive a simple anthropocentric formulation: Exploitation and depletion of biological resources will escalate so long as the *perceived* human benefits exceed the environmental costs.

Let us now consider what happens to the balance of human benefits and environmental costs when increasing numbers deny the option of mobility to subsistence agriculture and pastoralism. There are two possible case histories, depending on relative growth rates. In the first, we have already seen from the historical evidence of the developed world that when resource extraction greatly outpaces population growth, a demographic and economic transition is possible, resulting in greater intensification of agriculture, longer investment horizons, and greater productivity. Here, as the point of population stabilization is approached, food production greatly outstrips demand (Barr, 1981) and, were it not for the need to feed much of the rest of the world, the amount of land under cultivation would rapidly be on the decline. There is still some question about whether the long-term soil losses in, for example, the Wheat Belt, can be sustained (Brown, 1981), but undoubtedly the environmental damage per unit of production is far lower than it would be under sedentary subsistence farming. In the second case history, the environmental impact is substantially greater where the rate of resource production lags behind population growth and inhibits intensive farming practices. This situation leads to intensification of practices best suited to shifting agriculture and pastoralism. In consequence, the impact on both physical and biological resources is greatly increased and relief is sought in yet further expansion into marginal areas, creating a litany of environmental woes.

Development Strategies

We have seen that for the next few decades the world will be locked into sustained growth, with most of the expansion taking place in the developing countries, where the demographic and economic transition to a more developed status cannot precisely replicate the patterns of the past. What solutions can we derive from conservation principles to aid the transition and minimize environmental destruction?

In reviewing the options I fail to find any principles in conservation that are not incorporated into the development dicta of a multitude of disciplines and agencies already addressing the issue. The failure comes not in ignorance of the principles, but in their application and acceptance for reasons often beyond the reach of economic or political solutions. Rather than present a comprehensive outline of rural development principles, which I am ill qualified to do, I will consider in a simplistic manner both the routes by which balanced growth can be achieved and some of the techniques available.

There are basically four ways that growth and production can be brought into balance—slowing population growth, accelerating production, reducing environmental impact, and changing human perceptions. A conservation-based approach is meaningless without tackling all these simultaneously. The unifying principle from the conservation perspective is that a surplus of production over consumption can reduce environmental impact and hold the ultimate depletion of resources at lower levels by bringing about an earlier stabilization of population. We can look at each option in turn and consider what prospects it offers.

SLOWING POPULATION GROWTH. Growth in populations is achieved when birthrates exceed death rates; to stabilize numbers both must be matched. The present unprecedented growth rates stem in large part, as noted earlier, from a rapid drop in the death rate. The only realistic option available is to foster conditions that favor reduced births and greater investment in each offspring. Within a traditional subsistence society the benefits of additional labor will always outweigh the benefits of reducing family size, and it is for this reason that most population policies now recognize the limitations of promoting contraceptives in isolation from other incentives, although better availability and information on contraceptives does contribute more to population reduction than is widely appreciated

(Ejiogu, 1972). To overcome the social prejudices against birth control and suspicions of the motivation behind national contraception programs, emphasis is placed more on the benefits to each child, and the family, of spacing births (Ominde, 1972). Given that the population growth rate in LDCs is over twice what it was in the developed world during its rapid economic transition, lowering the absolute growth rate should be a prime goal of rural development.

ACCELERATING PRODUCTION. Increased resource productivity, especially food, has been a successful approach to creating a surplus and permitting other programs to take effect. This has been especially true in India, for example, where most of the gains in per capita calorie intake have been achieved by technical advances in production rather than a fall in the birthrate (Krishna, 1980). However, advances in the technical performance of grain production have tended to increase dependence on energy, fertilizers, and pesticides, which are economically less affordable and less available because of inflation over the last decade. Without technological breakthroughs, increasing productivity will depend on improving the effectiveness and availability of existing development procedures. That is not to deny the potential that exists already. The theoretical global maximum food production is some twenty-five times the 1977 output, assuming that all available resources and land could be used with optimal efficiency (Buringh, 1977). Though this may not be a desirable end point to human growth, it does show the potential if land, resources, and human capabilities can be marshaled effectively. New technologies could, of course, raise the global carrying capacity yet further.

There are, however, innumerable factors constraining the attainment of maximum production. High production often depends on monoculture, which involves an unrealistic risk for farmers with family and market commitments. Sustained production, in contrast to maximum production, requires the conservation of resources on which it depends—land, water, energy, and plant genetic potentiality. I will consider a few examples to emphasize resource conservation as the underpinning of sustained productivity.

FIG. 9.2. *Grain harvest in Nepal. In traditional societies, most of the energy exerted in cultivation is still from human and animal power. Photograph by M.J. Odell.*

Water. Crop production in many arid parts of the world is limited more by water availability than by nutrients; indeed, any dryland agriculture has first to guarantee water supplies before investing in other components of production. Crops with high water-use efficiency are essential in drylands, and some recent advances have been made in selecting better-adapted crop strains. Equally important are the methods of gathering, storing, and delivering water to crops. The ancient qanat method of subterranean water channels in the Mideast and Far East deserts is a superb example; it demonstrates that excellent transportation and water conservation can be achieved simultaneously, even in peasant communities. Modern techniques, such as drip irrigation, can reduce water requirements by 50% compared with furrow and flood irrigation practices. The drip method has the additional advantage of minimizing erosional and nutrient losses and of dispensing with land wastage for ditches.

In Israel the combination of two evils in crop production, sandy soils and saline water, is achieving success in desert areas. The saline waters, and in some cases seawater, give satisfactory yields of barley and sugar beet on permeable soils where the salts are not retained (Hanson, 1979).

Energy. Energy is essential for agricultural productivity and for future improvements in output. We saw earlier that the energy consumption per unit of output is three times greater in the developed than in the developing world. Is the energy–food production equation fixed, or can we improve efficiency and better utilize locally available resources? Because much of our energy base has so far depended on fossil fuels, new techniques of production and conservation are vital.

Most energy used in food production is consumed in land clearance, plowing, the application of fertilizers and pesticides, routine weeding, harvesting, processing, and transportation. In peasant agriculture, energy sources are at hand in the form of manpower, ox plows, dung, wood fuels, and the sun. The time necessary to apply such energy to crop production, however, prohibits intensification. Improved crops produced by genetic engineering could transfer many of the energy requirements from the farmer to the plant; thus, greater efficiency in photosynthesis and in water and nutrient use and better disease resistance could reduce the direct investment of the farmer. Other techniques for providing energy on the farm at limited cost are gradually becoming available. These include solar driers, wind generators, wind pumps, and the planting of crops such as *Leucaena*, which simultaneously provide high yields of firewood, serve as windbreaks, offer livestock fodder, and replenish soil nitrogen reserves. Biogas plants have also proved highly successful in China; with them the organic waste and compost used in gas production derive directly from farm refuse (van Buren, 1980).

Crop Potential. Choice of crops is particularly difficult for a peasant farmer shifting into a market economy, because he has limited knowledge of the stability of crop demand, prices, and so on. He is uncertain about the requirements, disease susceptibility, and compatibility with other plants of unfamiliar crops. New varieties of crops continually appear, and it is unclear in many cases what conditions favor their potentiality, what manpower requirements they have during growth, and how sensitive they are to disease, water deficiency, and nutrient status in the soil. Notwithstanding such problems, the potential presently available in crop varieties is sufficient to generate enormous gains in production and to diminish risk. Thus, various methods of multiple cropping—such as mixed cropping, intercropping, and rotational cropping—offer new techniques by which production can be increased without necessarily depending on either heavy fertilizer applications or pesticides, both of which are important considerations in emergent agricultural nations.

In the long run, greater potential yet may exist through selection for crop varieties tolerant of local resource deficiencies, and through genetic engineering for higher-yielding hybrids. Despite many problems of the green revolution, the consensus is that it has ultimately led to sustained higher food production in many developing countries, especially India and Pakistan, which have as a result more or less achieved self-sufficiency.

To date, new varieties have come from genetic selection over numerous plant generations; progress has thus been slow. Genetic engineering offers new opportunities and hope that the selection time can be reduced to a few years and that hybridization of previously impossible traits can be achieved. Quite apart from speeding up selection of higher-yielding strains through improved photosynthetic, nutrient, and water-use efficiency, there is the prospect that genetic engineering will permit crops to be structured to the environment in ways more efficient than were previously possible. Thus crops, and especially cereals, can potentially be selected for their capacity for nitrogen fixation, allopathy, and greater tolerance of environmental stress, such as salinity, temperature extremes, and drought. In other words, we are fast approaching a point where the plant can be structured to suit the environment, rather than the reverse. The potential for decreasing the dependency of Third World economies on imported agricultural requirements by these means is enormous.

REDUCING ENVIRONMENTAL IMPACT. Conservation of resources to reduce environmental impact is basic to sustained agriculture, and, with increasing land pressures and lack of opportunities to move elsewhere while the land recovers, there is a greater need now for applying the principles of sustained management than in any previous time. Let us examine a few examples that stress the significance of conservation.

Soils. Soils are essential for plant production; they are the substrate in which plants are anchored and from which they derive their water as well as organic and inorganic nutrient requirements. A good soil structure will make all the difference between stable, high yields and sporadic, ephemeral, and scanty production. The rate of soil genesis is rarely more than 2 to 5 tons per hectare per annum (Brown, 1981), yet in most agricultural areas erosional losses greatly exceed this amount. In many of the subsistence fields on the steep slopes of eastern Mount Kenya, for example, erosional losses exceed soil formation by two orders of magnitude. The prospects for any sustained agriculture under these circumstances are dismal. Conservation of soil structure and organic nutrients is essential in the arid zones, where the availability of water can be substantially reduced by loss of infiltration capacity, which in turn leads to greater overland flow and greater sediment and soil nutrient losses. Methods of soil conservation are widely understood but not necessarily widely practiced.

A recent method that offers hope of increasing plant production and decreasing the loss of both soil and nutrients is reduced or no-till agriculture (see "Examples of Conservation-based Development").

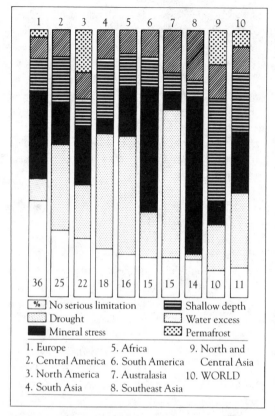

1	2	3	4	5	6	7	8	9	10
36	25	22	18	16	15	15	14	10	11

% No serious limitation		Shallow depth
Drought		Water excess
Mineral stress		Permafrost

1. Europe	5. Africa	9. North and
2. Central America	6. South America	Central Asia
3. North America	7. Australasia	10. WORLD
4. South Asia	8. Southeast Asia	

FIG. 9.3. *Regional distribution of soils with and without limitations for agriculture. Chart from IUCN, 1980.*

Water Conservation. It is difficult to isolate water conservation techniques from those that also improve production. Most conservation practices beyond the immediate farm level center on the catchment zone and involve flood control, watershed management, and so forth. Central government planning nowadays plays a large role in integrating the myriad considerations involved. The principles of water basin management, control, and conservation are well established, though often difficult to institute (Dunne and Leopold, 1978). Forest conservation also plays a crucial part in maintaining a year-round supply of water within a minimum of sediment loss.

Plant and Animal Resources. It was only a few decades ago that conservationists first argued that the genetic potential among living animal and plant species was an important enough reason to conserve as diverse an array as possible. Today that argument is taking on increasing significance with the realization that plant and animal breeding depends to a greater extent than ever before on the genetic variability and potential of both domestic and wild species. The International Rice Research Institute, for example, conducted an extensive worldwide search before finding resistance to grassy stunt virus among a single wild species. Today the fruits of that search are manifest internationally in the production gains of varieties crossed with the original wild species (Hanson, 1979). Centers for germ plasm have taken on renewed significance with the prospects of genetic engineering and have rapidly built up vast gene banks. However, even though such centers are vital for future research and development in plant sciences, they cannot store more than a limited array of wild species. The value of biological reserves as genetic storehouses of animal and plant resources has thus increased considerably, although the cost of their upkeep will have to be justified by additional uses. The need for minimum population sizes to offset the loss of genetic potential through inbreeding and chance hazards is still inadequately appreciated, but it argues strongly for more full understanding and utilization of principles of genetics and population dynamics in the design of biological reserves (Soule, 1980).

Recreational and Aesthetic Resources. With demographic and economic transition, recreational space requirements will inevitably increase. The European and North American examples point to the need to preserve adequate outdoor space for future generations in presently developing nations. History also argues for the need to separate land devoted to recreation from that allocated to wilderness and aesthetic enjoyment. With the exception of small city parks and sports fields, most large-scale, outdoor recreational and aesthetic activities require space that is only available in rural areas. Dams, reservoirs, and rivers can be opened to a variety of water sports, and timber reserves can double as recreational resorts. Areas farther afield with natural wilderness—including national parks—need to be set aside in advance of demand. However, given the short-term constraints on land, it may be necessary in some cases to allow interim, nondestructive uses, such as extensive pastoral ranching, on areas that can later be reconstituted as wilderness lands.

CHANGING HUMAN PERCEPTIONS. Recognizing the needs to slow population growth, accelerate production, and reduce environmental impact is the least of the problems in bringing about changes in rural development. The greatest challenge lies in implementation of these principles in the face of limited finances, infrastructure, trained personnel, and available resources.

Various institutions—forestry agencies, fisheries, wildlife departments, national parks, and a host of others—nowadays endeavor to put into practice plans to exploit natural resources on a sustainable basis. In some cases central planning and management is limited to stimulatory and regulatory actions to foster the efforts of private individuals and to govern environmental impact. This is normally the case in more developed nations, whereas in the developing nations growth and production, as well as environmental legislation, are more often centrally planned.

Whatever approach is taken, there is little question that the central intention is to change perceptions and stimulate the receptiveness of individuals to new methods, to make them aware of the long-term value of sustainable practices, and to bring about balanced growth. Concepts and techniques of balanced development are far ahead of our ability to put them into practice, and I consider their practical implementation the most important immediate challenge for the future.

I want now to present three examples that apply the use of these conservation principles in rural development. Each shows that innovations need not be based on new technologies to achieve immediate benefits in terms of rural development; in fact, they can often provide as much benefit simply by applying appropriate principles.

Examples of Conservation-Based Development

The first example deals with wood fuels, which account for most of the energy consumed in develop-ing countries. In Kenya over 75% of all energy used is derived from wood fuel. A recent survey (Western and Ssemakula, 1980) shows that the rate of wood consumption exceeds annual production by two and a half times and that the differential will widen as energy consumption per capita grows with increasing affluence. Urbanization further accelerates the reduction of wood mass, which in towns is used in the form of charcoal, thus delivering to the consumer less than 20% of the original calorific value of the tree. Projections to the year 2000 suggest that four-fifths of the country's woodlands could be stripped to provide fuel wood and make way for agricultural expansion. There are some constraints, however, that make this dire prognosis unlikely. These include the costs of transportation from distant sites, the limited biomass density of wood in the vast, arid lands, and the low bulk density (calorific value) and twig size of some trees, which make them uneconomic to harvest. As a result, pressures on wood stocks will be excessive in the agricultural areas, and extensive erosion could take place there. What solutions are available to alleviate the seemingly inevitable energy shortfall, given the costs of imported fossil fuels?

The simplest immediate solution is to improve the thermal efficiency of existing and widely used metal stoves, which are only half as efficient as cheaper clay stoves used in India, where biomass shortages have been widespread for decades (Oppenshaw, 1980). Further saving could be made by introducing charcoal kilns that are more efficient than the traditional ones. Improvements of up to threefold in the efficiency of energy transfer from tree to consumer could be obtained in this way.

The second example comes from experiments conducted at the Institute of Tropical Agriculture in Ibadan, Nigeria. There experiments tested the efficiency of no-till agriculture in a humid tropical region where various factors—including intense rainfall, which contributes to heavy erosion, rapid loss of fertility, and the considerable time that must be spent controlling weed invasion—constrain agricultural productivity under conventional tillage (Greenland, 1975; Wijewardene, 1978). Under the no-till method the total labor per hectare needed to raise a maize crop successfully was 2.3 hours, compared with 5.4 hours under conventional tillage. Soil loss was reduced from 1.2 per hectare to 0.03 tons per hectare on 1% slopes and from 23.6 to 0.14 tons per hectare on 15% slopes. And water runoff was reduced to one-fifth and one-fourth respectively on the same gradients. More significant, crop yields, which were similar in the first year, increased on no-till plots from a little under 3,000 kilograms per hectare to a maintained level of over 4,500 kilograms per hectare over the next three years of experiments. Conventional tillage increased to 4,000 kilograms per hectare in the second year and progressively declined to under 3,000 over the next two years; this is a falloff typical of shifting cultivation in the region.

In the same program simple devices were constructed to allow farmers to apply herbicides and fertilizers efficiently at low cost and to dispense seeds directly through the surface mulch. Only in the last decade have herbicides offered a practical alternative to conventional tillage. Although more research needs to be done on the biological effect of the herbicides, the evidence from this experiment is that the impact of low application rates, less than 0.5 kilogram per hectare of low-toxicity material, is minimal. The improvements in the soil's chemical and structural properties because of the persistent mulch resulted in a ten-fold increase in worm casts compared with conventionally tilled fields. In 1977 over 12% of the United States was planted using no-tillage agriculture, and a much larger proportion is expected in the coming two decades (Hanson, 1979).

The final example concerns wildlife conservation in Kenya, and an effort I have been involved with to have a national park contribute to the welfare of cattle pastoralists. Since 1947, when Amboseli was made a national reserve to protect its impressive wildlife populations, there have been attempts to establish a national park and remove the Masai pastoralists who traditionally used the area. The size of the ecosystem used annually by wildlife migrants is some 5,000 square kilometers, although a much smaller expanse, 600 square kilometers, serves as the dry-season concentration area around permanent water. Both wildlife and Masai used the same migratory routes and dry-season water sources (Western, 1976; Western and Henry, 1979).

Changes in the life-style of the Masai resulted in their demands for division of the area into ranches. No concessions were envisaged for wildlife that contributed no income to the pastoral Masai, yet consumed some 30% of their grazing annually. The local county council and government were, however, benefiting greatly from over 70,000 tourists annually, who contributed a total of more than $2 million to the area's economy. In an effort to accommodate both the interests of the Masai in the ecosystem and the future of wildlife, a core area of nearly 400 square kilometers was established as a national park to guarantee the survival of the crucial dry-season range. The Masai were given title to the remainder of the ecosystem for extensive ranching, and, to make up for their lost grazing and water in Amboseli, water was piped to areas usually too far afield to be grazed in the dry season. The Masai were also permitted access to important late-season grazing in the swamps around Amboseli. To encourage the ranchers to permit continued wildlife migrations, they were given an annual compensation fee to cover their grazing losses. In addition, the park's headquarters have recently been removed to the park's edge, where its communal facilities, such as hospital and school, are available to the Masai as well as park staff. Campsites have also been shifted onto the Masai side of the boundary, so that they can earn tourist revenues directly. Both moves alleviate the human pressure within the park while they benefit the inhabitants in the surrounding area.

Thus, the viability of the park has been gained by contributing directly to rural development; had this not been done, it is doubtful whether a national park would ever have been established. The lesson here is that revenue distribution can be more important than maximizing of returns. A few dollars received at the local level can create benefits out of all proportion to thousands received by individuals remote from the area.

The Challenge of Environmental Planning

The extrapolation of trends into the future rests on the assumption that existing interactions will prevail. Malthus, I feel sure, would not have predicted

FIG. 9.4. *A Masai herdsman and his cattle at a waterhole in Amboseli, Kenya. Photograph by R. C. Milne from IUCN.*

the inevitable crash he foresaw if he knew the human population would continue to grow for nearly three centuries more. Yet the Malthusian axiom that "the power of population is infinitely greater than the power in the earth to produce subsistence for man" is still incontestable. What has changed is our ability to produce subsistence, and with it has changed the time frame of Malthusian prognostications. The relevant question today is not so much what the ultimate capacity of earth is to support humanity, but rather what the costs are in terms of welfare and environment of supporting a given population. It is inconceivable that the human population will continue to increase with each technological advance until sheer congestion limits further expansion; indeed, the lessons of technological progress argue for a contrary trend, in which individuals are willing to forgo more offspring for a better standard of living.

Looking ahead, there is reason to believe that a concerted effort in rural development can improve the outlook for the future and conserve a greater proportion of the earth's natural resources than the

alternative of relying on the natural inertia of human growth to slow it. There is reason, I believe, to think that the turning point may already have been achieved. The 1950s marked the first time ever that agricultural production increased more by intensification than by expansion (Brown, 1981), and the 1970s signaled the first time in millennia that the rate of human population growth slowed. Moreover, the projected population increases reflect global figures and not necessarily those within the rural areas. Thus, although the earth's population may increase from 4.4 to 6.2 billion between 1980 and the end of the century, a quickening urban migration will substantially lessen the rural increase. If we assume that the urban migration rate will continue at 3.7% (Hauser and Gardner, 1980), compared with the global population growth rate of 2%, then of the additional 1.78 billion individuals added, only a quarter, or some 430 million, will be located in the rural areas. If we also accept the prediction that over 70% of the gains in agricultural output between now and the twenty-first century will be from intensification rather than range expansion, we have some encouragement that there is still room to maneuver and hope that conservation-based approaches can contribute to the long-term welfare of the environment and of our species.

Conservationists must contend with an era in which new land is even more limited for conservation than it is for development. We have moved beyond the stage when the major deficiencies in our conservation planning can be made good by adding on new land; in the years beyond the "more land" approach, it will be the resources we conserve within the rural areas, and the perceptions we have inculcated there, that will largely determine how much of the earth's present biological richness remains.

Discussion

QUESTION. Where did the solution at Amboseli come from? Was it something the Masai developed themselves, or was it something you or the government developed?

ANSWER. The government of Kenya could have come up with the solution we found at Amboseli if they had really thought about it. But, like most governments, they still think in terms of increasing the gross national product. This is true at least partially because they are trying to impress donor organizations. If their gross national product goes up, it makes a very good case for further funding. They might make slightly less money if they took a rural-based approach, and this would not be as attractive to donor organizations, even though it would offer more realistic and long-lasting solutions at the local level.

What we had in the Amboseli case was a compromise between the two approaches. The economists realized that who is benefited matters more than how much the total benefit is, so they advocated that the whole economy at Amboseli be directed back to the local level. Once that was set up as a philosophical foundation, most of the rest of the development came from the Masai themselves.

They were not organized in any sense; they are a people without a centralized government. They could not get together to make decisions given their state of development at that time. I operated very closely with them, trying to solicit their viewpoints. I said, "If wildlife is of value, if it does substitute for cattle in your economy, would you accept a solution that gave you more access to wildlife as the number of cattle decline?" This brought out strong feelings that had always been there. They said, "That is how it used to be, wildlife has always been our second cattle—until you people came along and took away the wildlife. What you are doing is suggesting that you put things back the way they were." This opened up a whole range of possibilities. Then the biggest problem became convincing the government that we had a realistic solution. This took many years. Finally, the way we worked around the government was to use the same economists to convince the World Bank that this was an instance where benefiting the rural sector was paramount. The World Bank saw this as a great opportunity. In turn, it told the Kenyan government they had to recompense local landowners for their losses as a precondition of a World Bank loan. You could say we used whatever means we had available.

QUESTION. In many cases people are up against the absolute limits of their resources, but in others it seems that they are only up against limits of their resources given the current distribution of wealth and their political systems. How much of this do you see as being a political problem, and how much is a real absolute shortage of resources? Is it distribution or availability?

ANSWER. In many agricultural countries the extent to which we are confronting an absolute limit is very minor. The Rockefeller Foundation and the West German government recently did a study of Southeast Asian farming showing that, with present technologies, many of the rice fields could produce about 22 tons per hectare. The current production only averages about 7 or 8. So a solution is not outside the limits of available production. Who is to say that 22 tons per hectare is itself the physical limit? With extra capital it could be increased.

In the other case I mentioned the limit is more absolute. The desert primitives, in particular, are at the present limits of animal husbandry, irrespective of the economic or political conditions. It does not matter how much we do to change conditions by massive inputs of money to provide water, because the animals do not alter physiologically. To select new strains that could resist those conditions would take generations. The knowledge even to attempt such a very-long-term program has not been available until now. For all intents and purposes, desert husbandry is at an absolute limit.

QUESTION. The lesson we learned in Southeast Asia was that if you increased production drastically, you also created tremendous political and economic effects. In many cases, while production did rise the net result was that small holders were driven off the land and larger, more advanced landowners took over. They then began producing rice as a cash crop, leaving large numbers of people who had been at a subsistence level now below it, starving. I think it is always important to remember those problems when you talk about increases in production. Would you react to this?

ANSWER. Let us say you introduce modern agriculture, and you reduce the manpower requirement by 90%. What happens to the people who are displaced? They become part of a social problem that is much more severe than the one you had before. The sort of situation you described is not only a function of the type of scheme introduced, but also of the indigenous situation.

In many areas there are problems because the pastoralists have traditionally split up the land into single family units. The areas are too arid for any one unit to have enough land to feed cattle for any length of time within these smaller areas. Now they have combined their holdings into much bigger units of land. There might be 40 or 50 or even 200 families moving around and using the same area. The pastoralists have accepted the new group strategy to some extent, but within that group, individuals want individual holdings. It is the way they have always divided the land. When they can afford to, they reject the new solution. Those few individuals who have acquired some capital are now trying to buy off others who have less useful land. Problems are not always imposed from outside, they are often internal.

We are going to see a very real trend toward urbanization in many of the developing countries. The problem is the same throughout the world. For example, the rate of urban development in Canada is 12% to 13% per year. The background population growth is only 4%. The concentration of landownership in a few hands is going to be one of the major political problems of the years to come.

QUESTION. Is the triage approach—forgetting the most desperate countries and concentrating aid on those that can be "saved"—something that will become prevalent among aid agencies? Will we be forced into it in the long run?

ANSWER. You have assumed that the donor country is going to be choosing where to give the aid. But the country that is soliciting aid also chooses its partner, depending on the type of commitment it can get and what strings will be attached. For example, if you look at the crescent of countries all the way from northern Uganda to southern Ethiopia, southern Sudan, and Somalia, they are all in the same predicament: They are all suffering a very severe drought. Which countries will the United States choose to help? A triage basis for making decisions will not help them in any sense whatsoever. The political mosaic will determine those who become its allies. Hence, Somalia is going to receive aid and Uganda will be ignored despite the fact that it has a more critical problem. Southern Sudan is already ignored at the government level simply because several American oil companies have begun working there and they will provide the "aid." It is difficult to see common ground for triage and political reality. However, the political solution may be equally inhumane.

Robert McNamara, who was head of the World Bank in the mid-1970s, once said, "Our aim as a World Bank is to abolish absolute poverty." Well, that all sounds good, but the World Bank is still a bank. In the last five or six years, the bank has not adhered to that principle for very simple reasons. They could invest literally millions of dollars in some of the desert areas, and the increase in production it would induce would be absolutely marginal. If they invested just a few dollars in a high-production area, they would foster a tremendous boost in production. The member countries of the bank themselves often define how much aid will be given and to whom. Then they select the donor agencies. This forces the World Bank, despite their principles, to operate on a far more pragmatic economic basis than they intended.

QUESTION. When the Masai took their own initiative toward finding a solution to their problem it worked. It seems that many problems are caused by outside parties imposing solutions or directive methodologies that do not coincide with the values or needs of the people. Am I correct in deducing from your comments that you are suspicious of any major governmental intervention?

ANSWER. I advocate that extension people, educators, and others who intervene understand the problems well and suggest alternative solutions. I do not advocate that they impose their own plans. In the pastoral case, I suggested that the best solution was for the government simply to try to find an appropriate market and forget all the sanctions they had imposed. Once the market is there, people will respond on their own. This is quite different in practice from traditional planning.

On the other hand, I do feel very strongly that a laissez-faire approach to environmental impact is completely counterproductive. If you say, "Let the people find their own solution to the lack of firewood," they will. They will cut down whole parks and leave only stumps. In such situations you must have a solution that contains safeguards. The protection of vital resources (and I would recommend a priority rating for determining which resources are vital) is one of the few cases where I think strong sanctions are necessary.

IV Priorities for International Action

Environmental Policy and Law

WOLFGANG BURHENNE

ALEXANDRE KISS

MALCOLM FORSTER

The World Conservation Strategy provides a good starting point for a discussion of international environmental law, but this topic should not be seen in isolation from national legal requirements. We will outline briefly the major reasons for an international legal approach to environmental policy and the mechanisms most frequently used. Finally, because prime responsibility for the realization of international environmental provisions lies at the national level, we will examine ways of increasing governmental effectiveness in conservation.

Reasons for an International Legal Approach

We must first consider why it is important to adopt a concerted international approach to environmental law. Four basic reasons are immediately apparent.

1. It is often said that the environment knows no frontier; air, water—both rivers and oceans—and wildlife cannot be contained by national borders. This is part of the message in the catch phrase of Stockholm, "Only One Earth."

In view of this fact, environmental conservation often requires international cooperation through law for two distinct cases: when problems center on certain shared areas (for example, river basins or border zones) or when issues concern mobile components of the natural environment (for example, migratory species, water, or air).

2. Economic factors provide another reason for cooperation through international law. Governmental measures for the environment usually impose charges on the national economy, at least in the short term. Generally this means disadvantages for exports, but it can also imply special restrictions on imported goods.

This dichotomy is very clearly discernible in the European Common Market directives related to environmental protection, more than sixty of which have been adopted so far. One of the most important, that of 4 May 1976, On Pollution Caused by Certain Dangerous Substances Discharged into the Aquatic Environment of the European Community, states in its preamble: "any disparity between the provisions on the discharge of certain dangerous substances into the aquatic environment already applicable or in preparation in the various Member States may create unequal conditions of competition and thus directly affect the functioning of the common market" (European Common Market, 1976).

Outside the harmonization of national legislation (which represents another means to avoid these nontariff barriers), governments and industry should seek to coordinate action on a multilateral basis. A parallel can be found in the international commodities market, where states often regulate supply and demand to conserve resources and in doing so affect market prices.

3. A third reason for an international legal approach to environmental matters concerns the conservation of natural resources located outside the jurisdiction of any state. It is in the interest of all states to cooperate to share equitably, use wisely, and/or conserve resources in the so-called global commons—for instance, the high seas or outer space. Should one state or a group of states seriously deplete resources in these areas, other states would be uniformly harmed. Thus, international environ-

mental law is needed to govern the exploitation of resources and other activities in such areas.

4. A final reason for international legal cooperation for environmental conservation is based on the acknowledgment of national responsibility to mankind as a whole for the conservation and/or management of exceptional natural areas or resources for scientific, cultural, ecological, or aesthetic reasons. A worldwide international convention on the protection of the world cultural and natural heritage, adopted by UNESCO on 16 November 1972, states that "parts of the cultural or natural heritage are of outstanding interest and therefore need to be preserved as part of the world heritage of mankind as a whole" (UNESCO, 1972).

This concept is not unfamiliar in the United States, where the idea of the sovereign holding of certain environmental values on "public trust" has been widely discussed in recent years. There are, indeed, earlier examples of international action to assist states in meeting a large obligation to conserve areas or resources of such importance through both financial and technical cooperation. Such cooperation has been more extensively used in the past for exceptional cultural sites—for example, the Abu Simbel Temples in Egypt—but we are now seeing more and more application for natural value, because man has the capacity to reproduce something manmade but he is powerless to recreate the natural world.

International Legal Mechanisms

The World Conservation Strategy focuses on international conventions and refers to the fact that a number of these deal directly with the management of living resources. The strategy does not tell us, however, that there are even a greater number of multilateral agreements concerning environmental matters, which consequently influence to a greater or lesser extent the conservation and management

FIG. 10.1. *Some natural areas such as Thompson Falls in Kenya are of such outstanding importance that they become part of the world heritage. Photograph by C. W. Leahy from the Massachusetts Audubon Society.*

of living resources. Some 300 such conventions are currently in force.

The strategy continues by focusing on four conventions of particular interest. But this initial focus should not obscure the fact that international environmental law is presently evolving in other directions as well, with a variety of mechanisms—some in the realm of "soft law," others a bit "harder,"—that influence international conduct. The strategy does not cover these areas of international environmental law well enough.

SOFT LAW. First of all, let us consider "soft law," which has particular importance in the development of international environmental law. A 1976 IUCN publication, *Survey of Current Developments in International Environmental Law*, divides nonmandatory texts (for instance, recommendations and declarations issued by intergovernmental organizations or international conferences) into three categories. All involve rules with which states should comply, but which cannot be traditionally enforced.

The first category of soft law identified is *directive recommendations*. These are nonbinding specific resolutions that propose action or codes of conduct for states. Such recommendations are frequently employed in international environmental law, for example, by the U.N. General Assembly and its specialized agencies, and by regional or restricted intergovernmental organizations, such as the Organization for Economic Cooperation and Development, the Council of Europe, and the Eastern European Council for Mutual Economic Assistance. Although the member states of these organizations are not bound to accept the recommendations, they cannot in good faith completely ignore them, and usually there is compliance.

The second type of nonbinding text is *programs of action*. In this category the most famous example is the Action Plan for the Human Environment developed at the 1972 Stockholm conference. Under these programs, an effort is made to define as clearly as possible objectives for international action and to determine the best means to realize them. Again, such documents are often acted upon. For example, the 1979 Migratory Species

Convention was developed in response to one of the provisions of the Stockholm Action Plan, and the European Economic Community is now drafting its third program of action on protection of the environment, its first two having been achieved.

The final category of environmental soft law covers *declarations of principles*. These are texts designed to form the basis of future international provisions for environmental conservation. The Stockholm conference again provides the leading example, in the form of the Stockholm Declaration, which enunciates twenty-six principles. A more recent example can be seen in the Charter for Nature, prepared by IUCN in response to a request from President Mobutu of Zaire. The charter was recently submitted to the U.N. General Assembly, which called for its further consideration.

STANDARD SETTING AND LICENSING. Another way international law for the environment has been developing recently is toward the creation of rules and obligations in the guise of international standard setting or licensing. In his contribution to the 1980 IUCN publication *Trends in Environmental Law*, Peter Sand describes this phenomenon as "transnational environmental law making."

Standard setting rules are rarely included in the basic text of an international treaty. Instead they appear in technical annexes and are subject to simplified procedures for adoption and amendment in response to scientific and technological changes, which are likely as our knowledge of the environment continues to progress rapidly. The standards may be mandatory or voluntary, depending on the legislative authority of the "mother" convention or arrangement.

Licensing arrangements are also increasingly popular. Certain activities are being made subject to special licensing requirements because of the risk they place on the environment. A growing tendency has been to develop international consensus, usually under a formal treaty arrangement, for the mutual recognition of national permits that observe certain common requirements. Examples of this may be seen in the waste disposal permits required under the 1972 London Convention on the Pre-

vention of Waste and Other Matter, the 1972 Oslo Convention for the Prevention of Marine Pollution by Dumping from Ships and Aircraft, the 1976 Barcelona Protocol for the Prevention of Pollution of the Mediterranean Sea by Dumping from Ships and Aircraft, and the 1976 Bonn Convention for the Protection of the Rhine against Chemical Pollution.

In both standard setting and licensing, the transnational rules developed are often subject to an auditing mechanism (for instance, national reports) to review governmental compliance. Such auditing plays a key role, for example, in decision making under the International Convention for the Regulation of Whaling.

CUSTOMARY LAW. A final source of international environmental law is customary law. Although most of the body of international environmental law has been the product of a rapid recent development—mainly through treaties, soft law declarations, and international rule making, as just discussed—we have also seen the creation of several important principles (such as Principles 21 and 22 from Stockholm) found over and over again in international texts. For instance, a series of recommendations has been adopted by the Organization for Economic Cooperation and Development; the most important for international environmental law were the 1975 Principles concerning Transfrontier Pollution. The U.N. Environment Programme has made an attempt to formulate the main principles that should govern conduct in the conservation and harmonious utilization of natural resources shared by two or more states. The draft approved by the UNEP Governing Council on 22 May 1978 has not been adopted so far by the U.N. General Assembly. However, some of these principles appear directly or indirectly in the U.N. Conference on the Law of the Sea Draft Convention; international treaties such as the 1979 Convention on Long-Range Transboundary Air Pollution, which was signed by thirty-five states—nearly all European countries, the United States, and Canada— and treaties concerning the protection of rivers or lakes against pollution.

FIG. 10.2. *An oil control boom in Casco Bay, off the coast of Maine, exemplifies industries' response to the well-developed body of formal environmental law on marine pollution. Photograph from Massachusetts Audubon Society.*

Principles that might ultimately be considered part of international customary law for the environment include

A recognition of sovereignty over natural resources

States' responsibilities to ensure that activities within their national jurisdictions or under national control do not cause damage to the environment of other states or of areas beyond the limits of national jurisdictions

The principle of international cooperation in the use of shared resources

Assessment of the foreseeable effects of activities that might significantly harm the environment of other states and areas outside any national jurisdiction

The duty of sharing information and holding consultations with other states in cases of possible transboundary harm

Cooperation in environmental emergencies

Equivalent access to and treatment in judicial and administrative proceedings for nonresidents

Nondiscrimination in the application of national legislation to activities that may deteriorate the environment inside or outside national jurisdiction.

Still, much needs to be done. For example, "social obligation of property" in a global sense for the proper sustainable use of natural resources should be recognized.

Increased Governmental Effectiveness

To return to the World Conservation Strategy, Chapter 15 concludes with a plea for increasing development assistance to conserve and make the best use of natural resources. There is not much to add in this overview to what is written there. We should stress, however, that aid agencies must be made aware of the importance of assisting the development of legislation and institutions for environmental conservation in recipient countries. Such supporting measures are too often ignored, but without them the lasting value of development assistance projects is doubtful.

Finally, we would like to address briefly two topics not sufficiently considered in the strategy, the "implementation gap" and the political aspects of governmental organization. A major problem we have to face in environmental law is inadequate enforcement. At an international level too many treaties are not in force because they have not attracted the requisite number of ratifications. At another level, we often have excellent laws on paper that are meaningless or very weak in practical application. The strategy touches on the difficulties developing countries sometimes have in enforcing conservation legislation because of social or economic realities. This is, of course, a very real problem. But what the strategy does not make clear enough is that enforcement may also be a problem in the developed world for different reasons, including overlapping competences, lack of personnel, legal texts so complicated that enforcement officers have difficulty grasping all their aspects, and, finally, in many instances a cavalier public attitude toward environmental regulations.

The approaches to this implementation gap will necessarily differ in the developing and developed worlds. But in both cases a crucial element is governmental will to make environmental measures meaningful. Regardless of the level of development or political system, sound national legislation is essential for conserving the natural environment. Not to implement such legislation might be considered a failure in responsibility to—or even a crime against—both present and future generations. Thus, all of us involved in environmental law have an urgent duty to devote attention to the causes and resolution of the implementation gap in all parts of the world.

We have some perhaps unorthodox views about requirements for governmental organization and the functioning of international bodies. Contrary to the basic conceptions of the World Conservation Strategy, we believe that particular arrangements can be less important than the personalities of agency leaders. This may seem simplistic, but unless there are good active people in important governmental or intergovernmental positions, nothing happens in administration—even if the governmental or international structure is sound. Indeed, it is impossible to formulate a textbook description of an optimal national or international structure because of the many variables that apply.

The role politics plays in the selection of government organization structures cannot be neglected. For example, in recent years the mandate and role of the ministry responsible for environmental matters in France have been changed several times; this happens everywhere fairly frequently, basically because of political considerations—the creation of cabinet coalitions or even sometimes bestowal of political favors. With such forces—which are perfectly understandable—at work it becomes a bit academic to place too much stress on optimum requirements within a government structure. It is essential, however, that the needed functions be fulfilled and that effective coordination mechanisms exist.

Discussion

QUESTION. Do the vast differences in various countries' legal systems leave you pessimistic about the possibility of designing international environmental laws that are compatible with all of them at once?

ANSWER. Despite all the differences, the similarities are greater. The terminology may be different—so may the division of labor, the delegation of governmental responsibility, and the derivation of legal concepts—but basically environmental law in each country is the same.

First, there is the same format. There are usually local jurisdictions, some intermediate governmental levels, and a federal jurisdiction. Sometimes the powers of the state and local governments are delegated downward from the federal; sometimes they are delegated upward. In either case, the results are the same.

Second, there usually is a body of statutes creating regulatory law, which is administered by specific governmental agencies. This new law almost never supersedes what we call common law, the law of custom or the law of court decisions. This mixture of statute and common law means that both agencies and citizens have duties and rights that are enforceable in court.

Third, the law that is called environmental law in each country does bind the persons, corporations, and governmental agencies that have to be bound if we are going to solve environmental problems. Every country has some sovereign authority and sovereign obligation to manage its own resources, such as land, important species, and navigable water. Every country has police power—the power to govern the conduct of persons, corporations, and agencies. Every one has a taxing power and a spending power to make good its obligation to fund the agencies it creates.

Still, I am somewhat pessimistic about international environmental law, because we have learned some sad lessons at the national level in every country. Pollution does not respect boundaries, national or international. It demands a cooperative response. However, that response is only now beginning. It will be necessary for all of us to push our national leaders toward stronger international cooperation.

QUESTION. As a lawyer who manages specific cases in court, I have a hard time relating the international conventions you discuss to my personal day-to-day experience. International law is more a statement of principles that relies on the force of persuasion than it is a set of regulations. When we try to protect wetlands and other critical areas, we have to use local laws, bylaws, or regulations administered by local communities and state agencies. We are dealing very close to home, trying to think globally and act locally. This is the only way to address many of the problems raised.

One question I ask myself is whether an international law or convention creates a right or a duty. If it does, it can be enforced by somebody; if not, there is little you can do actually to enforce it. Is it useful even to mention documents unless they create an obligation on the part of a country, an individual, or a corporation?

ANSWER. To some extent, your analysis is correct. For example, I insisted on a critique of the wetlands

convention in the World Conservation Strategy. It really creates no obligation of any sort on states or citizens. It is a broad statement of concern. Do you get the impression that wetlands are protected by this convention? They are not. There have been attempts to give the convention some teeth, but there has not been any official move to amend the treaty, and perhaps this is deliberate.

But now consider the Natural Habitats Convention. This is a completely different type of document. It was first drafted by UNESCO to preserve the world's cultural heritage. Under pressure from IUCN, examples of natural heritage were added at the last moment to the list of features to be protected. All major states are contributing to a fund that will help countries possessing areas of international importance to maintain them. This is not a treaty to help the United States, or Germany, or the United Kingdom. It will help maintain important natural areas in countries that do not have the means or ability to maintain them alone.

The third example I would like to mention is the Convention on Trade in Endangered Species. It has requirements that bind not only states but also the individual citizens of these states. You cannot buy products from endangered species, you cannot sell them, and you cannot export or import them. It is a very tough treaty. As an aside, I should mention that its enforcement is threatened in the United States at the moment.

The range of legal force is exceptionally broad in international agreements, but even the weakest serve some purpose. Penalties aside, it is embarrassing for a country to be caught violating principles it has promised to respect.

QUESTION. Even if there is a sudden surge of acceptance and support, I am not sure that the present international process can successfully handle global environmental problems. The mechanism is simply too cumbersome. Do you think we have the system we need to manage global problems?

ANSWER. In the short term, I think we have to live under the conditions that exist. International treaties to establish new environmental responsibilities are the only possibility we have right now. An international management agency would be an easier way to handle many problems, but I do not think there is a chance for this at the moment.

Consider the way the ozone layer will be managed. Sweden has now promised to take the lead on this problem in the international community. If enough states agree to participate, Sweden will host a high-level meeting on environmental law, which will lead to a treaty. Only if the states become members of the treaty and enforce it in their own countries can we succeed in managing the problem. This sort of system does not promise any great degree of success.

In the long run environmental megaproblems will be beyond the capacity of nation-states to manage this way. They are not going to be solved nationally, or even internationally, under the present system. We are now at a point where asking about other alternatives is necessary. Environmental law finds itself at an embarrassing crossroad. There are quite a number of recognized problems. They have been tackled on an international plane, in some cases as far as they can usefully be addressed on that level. In the short term it is now necessary for national bodies to do something about these issues.

What concerns me are the long-term problems. There may be problems infinitely more serious than some of the ones we have been dealing with, problems about which we know absolutely nothing—major exploration on the deep seabed, disruption of the benthic community, and who knows what else. The first priority is to know what is going on. We will need massive research programs on these subjects. There is no hope of getting such programs at the international level given the current system. And there is next to no hope of getting them nationally. Deciding how to conduct joint research and joint management must become the preeminent challenge for environmentalists and for the international community. I am afraid they are not going to measure up.

QUESTION. Do the designers of international environmental law even know that we are out here? It seems that there is an enormous distance between the international legal community and the vast majority of environmentalists. Perhaps closer cooperation would bring better results.

ANSWER. How many international environmental treaties does your Senate ratify? How many are promulgated by the English or French parliaments? Not many. One of the most important local actions is to give a strong spur to national governments. They are the real impediments to progress. Very often environmental legislation is ignored because it does not seem to have popular support.

You must also try to gain the support of your individual leaders. They are often more important to success or failure than the institutions or the laws they represent. Any agency depends on the person at the top to set its direction. You have to be sure that person is being encouraged to work in an environmentally conscious way.

Individuals must realize how important their own input is, especially if there are equally aware people in other countries. To a remarkable degree, what happens at the international level in the next decade will be a result of people in one country taking as their own the interest of people in another. Of course, this sort of system works best when it is reciprocal. People must work nationally and locally for mutual goals. Much of the progress that is being made comes from such efforts. It is always important to remember the effect of people acting in private roles, on conservation commissions and so forth. All the more formal trappings will follow as a direct consequence of the work these people do.

There is an intimate connection between the implementation of international law on the one hand and the nature of the relationship between states and their individual citizens on the other. Despite the progress that has been made toward an international body of some legal authority, in the short term we need another strategy. We need to educate people to take individual responsibility for the environment. These people will generate a sense of responsibility in the international community and in their national governments. Real-world political action starts with individuals who insist that their governments act responsibly.

Management of
The Global Commons

ANN L. HOLLICK

The global commons is a form of common property resource. A common property resource is available to multiple users without exclusion, that is, its use by one nation does not preclude use by another (Hardin, 1968). The global commons include the oceans, the Antarctic, outer space, and the shared environment. Because the actions of one nation with regard to a commons resource may affect other nations, management of the global commons is an ongoing foreign policy problem (see *Journal of International Affairs*, 1977).

Certain historic patterns of development have characterized global as well as other common property resources. In the first stage, technological change makes new areas or opportunities available and, in effect, creates new resources. The second stage is one of active exploitation of the newly available resource. Because of demand generated by the rapid expansion of the world's population and income, a number of commons resources have been subjected to overexploitation and in some cases depletion. Cases in point include the periodic depletion of Europe's fisheries or certain types of whales as well as overuse of land, rivers, or atmosphere in certain parts of the world for disposal of waste materials. Some commons resources, on the other hand, are at an early stage of exploration. These include Antarctic and deep seabed minerals. At various stages of development between the two extremes are Antarctic krill, offshore oil and gas, geostationary and elliptical orbits for space satellites, the radio spectrum, and some aspects of the atmosphere.

Management Options

Three options exist to deal with global commons

resources: laissez-faire, exclusive management via partition, or joint management by more than one nation (Brown et al., 1977). Of the three approaches, laissez-faire has been the most practiced, and in many cases has resulted in problems such as overexploitation. The classic cases are Europe's offshore fisheries and worldwide whaling. The Second World War temporarily precluded European exploitation and allowed these resources to revive. The sardines off the shores of California, on the other hand, became extinct. In another case—heavy atmospheric or river pollution in Europe—national and regional measures have relieved the problem. The 1967 Outer Space Treaty and the 1958 Geneva Convention on the High Seas are laissez-faire in their approaches, despite the fact that they call for due regard for other users and set general principles for activities.

The second management option—partition—has been employed in many domestic commons. Recently the international community has agreed to the partition of offshore resources. Ocean areas that were formerly part of the ocean commons have been carved into 200-mile economic zones belonging to the coastal states. This represents an allocation of property rights to nation-states on the basis of some concept of geographic rights derived from proximity or historic claims. The possible partition of the Antarctic is in a state of suspension. Another forestalled effort at partition is represented by the Bogotá Declaration of 1976, in which eight of the eleven equatorial countries laid claims to airspace above their countries including the geostationary orbit.

The third approach to managing a global commons—joint management—may make the most sense economically for some resources and areas.[1] Joint management does not necessarily require the establishment of supranational international organizations, as called for by functionalism. It is nonetheless politically difficult. Joint management approaches encompass a range of possible schemes from consultation, liability, and regulatory schemes at one extreme to comprehensive management and

centralized resource allocation at the other. In the 1972 Convention on Antarctic Seals and the 1980 Antarctic Living Marine Resources Agreement, joint management approaches are being tested. The Mediterranean Action Program is developing a mixed system of regulation and liability. The World Administrative Radio Conference has evolved toward a regime that in some regions calls for preplanned allocation of the orbit and frequency spectrum resources. The recently adopted Law of the Sea Treaty seeks to resolve disputed property rights by creating a very elaborate system for centralized management and allocation of seabed resources.

Choosing among these three management options raises a number of difficult political, institutional, and economic questions. In fact, the choice is always made, either deliberately or by default. The political questions are essentially equity issues. In determining which approach to pursue, decisions must be made about who will participate and how, how decisions will be made, and who should benefit.

The overriding institutional question is how issues should be grouped or linked. For example, should all ocean issues be dealt with in a single U.N. conference on the law of the sea? Or should fisheries management be linked to other food regimes and seabed minerals issues be linked to Antarctic and other land-based minerals regimes? These types of decisions are typically made on political rather than scientific or technical grounds, as one group perceives a negotiating advantage in one or another form of linkage.

The economic questions of how to exploit and maintain a resource most efficiently are generally among the last to be addressed. Each resource often has physical and technological attributes that determine how the fewest expenditures (of capital and labor) can be made to recover it. Clearly, the creation of an elaborate and costly seabed authority meets political rather than economic or efficiency goals. On the other hand, partition of the continental margins among coastal states allows each government to extend its existing national regulatory framework to the new areas. In this context, the resources used to develop each oil pool can be judiciously limited if the government in question is

[1] An early example is the Fur Seal Treaty of 1911 for the Pribilof Islands between Canada, Japan, and the United States.

concerned with efficiency. The partition approach has worked less well, however, for the fisheries above the continental margins. The United States, for example, is creating a monopoly for domestic fishermen based on hidden and overt subsidies and an expensive management scheme. Foreign fishermen are being excluded from access to all the surplus catch. Foreign processing may also be excluded. The result may be to reduce the welfare of the U.S. taxpayer, who is paying for the higher-cost fish.

In the cases of living resources as well as the marine environment, the actions of neighboring coastal states in managing their 200-mile zones could affect one another's resources. As with the atmosphere, the movement of ocean waters generates certain unavoidable interdependencies. The lesson is that the approaches to common property resources that achieve socially efficient solutions will vary depending on the attributes of each resource.

Ocean Resources

Among the global commons, the oceans have the longest history of exploitation and international negotiation (Hollick, 1979). The Third U.N. Conference on the Law of the Sea (UNCLOS III) was only the latest of a long series of attempts to codify the rules governing exploitation of ocean resources. Through a combination of customary law and successive codification efforts, the law of the sea has evolved toward a mixed management regime—partition of coastal resources and proposed joint management of the seabed mineral resources beyond 200 miles. Figures 11.1 (fisheries) and 11.2 (continental margins) illustrate the reasons behind the impetus for partition. As the maps indicate, the ocean resources of prime economic value are located in coastal waters or on the continental margins. Let us look at the management of each ocean resource.

FISHING. Fishing, together with navigation, is a historic ocean activity. Most fishermen have operated in near-shore areas. As the distribution of phytoplankton shown in Figure 11.1 indicates, the major fisheries are found within 200 miles of shore.

The map also shows that countries with the longest coastlines are the major beneficiaries of offshore zones. Although the 200-mile-zone concept originated in Latin America, the developed countries benefit substantially from it (see Hollick, 1981, pp. 75–80). One-fifth of world fisheries, for example, lie off U.S. and Canadian shores. On the other hand, Asian and African states often have short coastlines or offshore areas that are abutted by other states.

Despite the geographic inequities, the UNCLOS adopted the partition approach to fisheries management. The problem of equity was solved (or ignored) by simply moving national boundaries out to sea. Each country is now called on to manage the living resources within its "national zone." Although this approach may be viable for countries with long coastlines fronting on open ocean, it will eventually have to be modified where the fisheries of a region are ecologically interlinked across national boundaries. On the coasts of West Africa and Southeast Asia the activities of one nation in its zone will affect its neighbors. Thus, although extending national jurisdiction is a principal preoccupation at the moment, further down the road coastal nations will have to develop cooperative approaches to manage coastal fisheries.

Two other types of high-value fisheries pose special management problems—tuna and salmon. Tuna is a highly migratory fish and moves at great speed both within and beyond 200 miles of shore. The fact that tuna migrate across national zones has led to conflicts between distant water and coastal fishing nations. The UNCLOS III Treaty sought, unsuccessfully, to grapple with tuna management. It reflects the developing countries' preference for controlling tuna on a zone-by-zone basis. However, in the long run, survival of the fishery will require a cooperative approach to managing tuna throughout its migratory range.

The behavior of salmon is equally problematic. Salmon is an anadromous species, which means that it spawns in fresh water and travels far out to sea. Beyond 200 miles, salmon may be subjected to overfishing that could undercut all the efforts of a coastal state to maintain the spawning streams. Therefore, as in the case of tuna, salmon fishery management must be undertaken on a cooperative

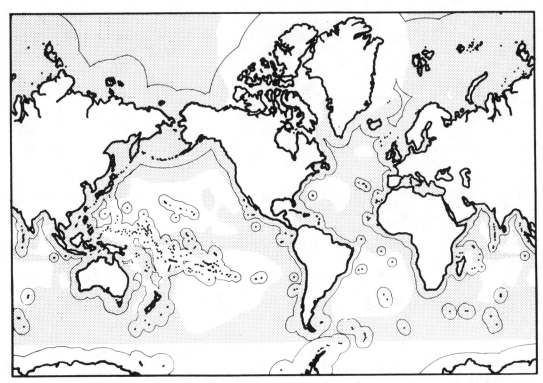

FIG. 11.1. *Phytoplankton Production. Areas of greatest phytoplankton production, which also have the greatest fish production, are shaded. Two-hundred-mile zones are shown by light lines.*

basis unless all salmon fishing is restricted to streams of origin.

Cooperative management approaches are likely to develop slowly at best. Most often cooperation will be forced by adversity. Only after overfishing or pollution have caused negative transboundary impacts will nations come reluctantly to accept and act on their mutual vulnerability and interdependence. Cooperative approaches are also politically difficult, so semicrisis conditions will be needed to prod their adoption.

NAVIGATION. Navigation, like fishing, is a historic use of the oceans. It has developed, over the centuries, a regime based on customary practice and treaty law (for example, the 1958 Geneva Conventions on the Territorial Sea and the High Seas). Because commercial navigation spans the globe, regulatory regimes must be developed on a worldwide basis. All countries engage in international trade, so all share an interest in avoiding piecemeal environmental and regulatory schemes.

For this reason, UNCLOS III was not the forum for serious efforts to develop international regulatory and liability regimes. Coastal states claiming "archipelagic" or "straits states" status have insisted on their right to determine shipping lanes. For the most part, however, the Intergovernmental Maritime Organization will retain responsibility for developing navigation codes of conduct. By the end of the century, it will have to come to terms with the need to control traffic in highly congested areas. Straits states will play a major role, but this issue should be handled by the international community that uses or has a stake in passage through these areas.

OFFSHORE OIL AND GAS. In the cases of offshore oil and gas, partition is a viable management option as long as two problems are properly handled:

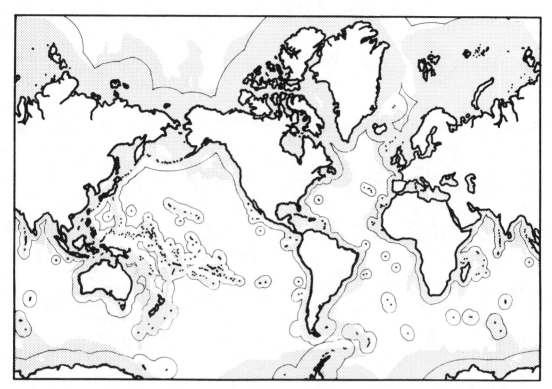

FIG. 11.2. *Continental Margins. Most oceanic mineral exploration in the near future will be limited to the continental margins. Continental margins are shaded; 200-mile zones are shown by light lines.*

(1) recovery from a common pool that spans national boundaries; and (2) attention to pollution impacts on adjacent states. As our capabilities have allowed us to drill in and recover hydrocarbons from greater depths, offshore oil and gas have begun to provide almost one-third of the world's oil supply. The prevailing expectation is that offshore hydrocarbons will be limited to the continental margins (see the shaded areas in Figure 11.2). There is also likely to be a correlation between onshore and offshore deposits. Thus, a division of marine oil and gas resources among the coastal states is not necessarily the most equitable approach to allocation. That is, countries that already have substantial onshore resources—such as Mexico, the United States, and Saudi Arabia—will get more offshore. However, in UNCLOS III, political expediency dictated that coastal states would divide these resources among themselves.

Partition of the continental margins among coastal states is likely to be the most economically efficient way to recover oil and gas. When a single state has clearly defined property rights in its offshore deposits, it will encourage its operators to recover these deposits with a minimum investment of labor and capital. Thus, the coastal state could ensure that, in the development of each pool of oil and gas, appropriate factor inputs and rates of recovery would be used. Difficulties arise, however, if an oil or gas deposit crosses a national border. As we have seen in the North Sea, states may engage in competitive drilling to take the most oil out of the pool before their neighbors do. The initial pattern of Norwegian drilling, for instance, was only along its boundary with the United Kingdom. It would be better to manage the common pool jointly to limit factor inputs and share recovery costs and benefits.

Similarly, states must be attentive to the environmental impact on their neighbors of their drilling activities. While one state may wish to concentrate on a rapid exploitation of its offshore resources, a neighbor may prefer to focus on tourism, recreation, fishing, and other activities that require preservation of the marine environment. It may, therefore, adopt more stringent safety and environmental measures. A case in point was the blowout of Mexico's Ixtoc well in the Bay of Campeche. In this situation, the United States found itself at the mercy of Mexican safety and environmental policies and procedures, and the state of Texas suffered substantial financial losses in tourism and fishing. Because of the strong emphasis on the rights rather than the responsibilities of coastal states at UNCLOS III, there are no provisions in the convention that would serve as bases for legal action or cooperative management. However, the U.N. Environment Program regional seas program is developing action strategies to fill the void. And as the pace of offshore activities increases, more and more incidents will underscore the mutual vulnerability of coastal states, and regional and global principles will be developed to address the question.

POLLUTION ISSUES. Pollution issues also arose at UNCLOS III with regard to shipping. And, as with resource exploitation in national zones, the conference was torn between contradictory impulses. In the case of shipping, coastal states have clearly wanted to protect themselves from pollution caused by vessels transiting near their shores. On the other hand, all nations perceive an interest in ocean shipping, whether they are shipping nations or are simply dependent on oceanborne trade. The conference sought to avoid a piecemeal approach to regulation of shipping through national or economic zones. Instead the UNCLOS III text seeks to balance coastal states' environmental concerns with the general interest in navigation by maintaining the principle of flag state enforcement. The coastal state may lodge complaints with the state where a vessel is headed. That port state may then take the appropriate verification measures and bring action against the offending vessel if the charges are deemed valid. However, flag state enforcement rights against violations can supersede all other legal actions under way where the flag

state has developed a credible record of punishing safety and environmental rule infractions.

DEEP-SEABED MINING. In the oceans context, the major exception to the partition approach was the negotiation of a comprehensive management regime for the deep seabed. The resources of principal interest in the seabed are manganese nodules and the recently discovered polymetallic sulfides. Nodules of greatest commercial value (those with the highest copper, nickel, and cobalt content) are found in the deepest part of the ocean, largely in the Pacific Ocean well beyond the limits of national jurisdiction. Because of early visions of vast mineral wealth, the U.N. General Assembly passed a resolution in 1970 declaring the seabeds beyond national jurisdiction to be the "common heritage of mankind." The most notable impact of that resolution was that most coastal states immediately laid claim to 200-mile zones and to the continental margins where they extended beyond 200 miles. Coastal states thereby succeeded in removing most valuable commons resources from prospective international management.

In subsequent negotiations on deep-sea mining, the mining industry and governments of the developed countries recommended an economically efficient approach based on a minimum of regulation. This would in essence allow miners to explore and lay claim to seabed sites. Conflict with other miners would be resolved through dispute settlement procedures developed among the mining states. Revenues from resource exploitation would be shared with the international community in acknowledgment of the much-disputed common heritage principle. Developing countries, on the other hand, have interpreted the common heritage concept to mean something quite different. Lacking the technology and capability to mine the seabed directly, the Group of 77 proposed that a seabed mining body called the "Enterprise" be set up under the auspices of an International Seabed Authority (ISA), which they would control. The Enterprise would mine the seabed directly to the exclusion of all national mining.

The UNCLOS III Treaty sought a compromise between the positions of the developed and developing countries. It was based on the concept of parallel tracks, or the "parallel system" as it has come to be called. On one track, private or state companies would be able to mine part of the seabed if they could fulfill certain conditions that would allow the Enterprise to go into business at the same time. These conditions begin with the requirement that each company explore and submit to the ISA the coordinates for two sites of equal value. The authority would select one for the Enterprise. Other requirements would include substantial up-front contributions to the ISA by governments, limits on production to protect land-based mineral producers, and mandatory transfer of technology to the Enterprise and to developing countries if the mining technology is not available on the open market. The parallel system satisfied neither side completely. In the view of many of the developed nations, it imposed onerous and economically costly burdens and set unacceptable political and economic precedents. Unable to negotiate needed changes in the seabed part of the Law of the Sea Treaty, the United States finally voted against adoption.

The future course of customary law for seabed mining remains uncertain. While the developing countries insist that an international seabed authority be created along the lines of the treaty, they do not have the resources or the technology to carry this out without the participation of the technologically advanced nations. For their part, some developed countries may prefer to establish a barebones development-oriented approach to seabed mining by banding together to resolve disputes and develop a legal framework for mining apart from the UNCLOS III Treaty. Eight developed states have negotiated a reciprocating states agreement that would allow them to recognize the mine sites claimed by others. The international consortia involved have negotiated among themselves to resolve the problem of site overlap.

In summary, then, there are a variety of approaches to the management of ocean resources. In the cases of fisheries and offshore oil, the partition approach has prevailed. Navigation and environmental resources have been left to existing international regulatory institutions, with some increased rights for coastal states. And seabed mining has been treated with a comprehensive management approach. In general, the approach to the oceans problems has been to promote national control through partition where possible. When they are politically necessary, namely in the case of the deep seabed, joint management approaches have been considered. In specific regions, such as the Mediterranean or the Baltic, cooperative management efforts are growing out of near-crisis situations in fisheries and pollution. Those who are interested in promoting such multilateral approaches should be alert for similar opportunities where physical realities impose themselves on policymakers.

Antarctic Resources

The Antarctic and its resources are at an earlier stage of resource exploitation than the oceans. This is in part because of the difficult physical environment and in part because of the complex political regime that governs the Antarctic (see CIA, 1978). The general approach has been to respond to resource issues as they develop on a joint or cooperative basis, but with an emphasis on national enforcement and self-policing.

Antarctic resources are essentially governed by a sixteen-member Antarctic Consultative Group. This arrangement has developed out of the political and scientific history of the area. In 1957 and 1958 the International Geophysical Year included an Antarctic research phase. The twelve countries participating in Antarctic research included seven states that had claimed parts of the Antarctic (Argentina, Australia, Chile, France, New Zealand, Norway, and the United Kingdom) and five countries that had not (Belgium, Japan, South Africa, the United States, and the Soviet Union).

In 1959 the twelve researching governments decided to formalize their cooperative scientific relationship in a treaty. The Antarctic Treaty applied to the area below 60 degrees south latitude. It provided that research plans and findings should be freely exchanged. The area was only to be used for peaceful purposes, and treaty state observers had free access to and inspection rights for all installations and equipment. The treaty further provided for peaceful dispute settlement and for an ongoing consultative mechanism. New members could be-

FIG. 11.3. *The Antarctic is quickly becoming an area where conservation and development will come into conflict. Photograph by D. S. Boyer from the National Geographic Society.*

come voting members of the consultative group if they undertook substantial research in the Antarctic. Because scientific research in the polar regions is an expensive undertaking, only four nations—Poland (in 1977), West Germany (in 1981), India (in 1983), and Brazil (in 1983)—have joined the group. Other states have, however, acceded to the convention; they include Czechoslovakia, Denmark, East Germany, the Netherlands, and Rumania.

As resource issues have arisen, the consultative parties have handled them on a case-by-case basis (see *University of Miami Law Review*, 1978). Two types of resources have been at issue—living and nonliving. The basic political issues have been those of claimants versus nonclaimants and, where claims overlap, claimants versus claimants. An important factor in overcoming these tensions within the consultative group has been the eagerness of the United Nations to have a role in managing the Antarctic. The prospect of U.N. involvement has been a significant incentive to maintain the cohesion and viability of the Antarctic consultative process. At the 1983 General Assembly, the United Nations resolved to study the Antarctic issue.

The question of regulating the take of Antarctic whales has existed since the 1930s. The International Whaling Commission regulates whales on a global basis because of their migratory range. The consultative group has focused on whale resources

that are limited to the Antarctic. Consultative meetings are held at two-year intervals, and at these meetings recommendations on a variety of scientific and resource issues are adopted. In 1964 Agreed Measures for Conservation of Antarctic Fauna and Flora (Recommendation III-8) was adopted, and in 1966 the Interim Guide Lines for the Voluntary Regulation of Antarctic Pelagic Sealing (Recommendation IV-21) went into effect. The latter served as the basis for the 1972 Convention on Antarctic Seals, which was negotiated in a larger framework that reflected the wide migratory range of seals.

More recently, the displacement of Soviet and Japanese distant-water fleets from the newly claimed 200-mile coastal zones has put new pressure on Antarctic living resources, in particular krill. In 1980, after years of difficult negotiations, the Antarctic Consultative Group finalized the Antarctic Living Marine Resources Agreement; it came into effect in 1982. The problem of how to deal with the seven claimants was handled by expanding the resource management area to include an ecological unit known as the "convergence zone." This zone contains all interrelated Antarctic species, and it contracts and expands with seasonal change. From the political perspective, the zone is important because it encompasses islands that are under recognized national jurisdiction. Thus, it was possible to develop a formula that begged the issue of national claims to the Antarctic mainland by referring to recognized areas of national jurisdiction. The treaty relies on data collection by participating states to keep informed about the condition of the resource. Assuming that all participating scientists agree that the data suggest a need to cut back on the level of fishing, consensus decision making is required to effect a system of allocation. Clearly, it will be difficult first to achieve agreement on the scientific evidence and second to take action if the evidence indicates the need for a reduced fishing effort. At least, however, the 1980 agreement takes the initial steps in laying the foundation for cooperative management.

The nonliving resources of the Antarctic, particularly offshore oil and gas, have been of increasing interest in the last decade. Little is known of the continent's mineral potential, although the Antarctic is expected to contain minerals similar to those found in Australia, South Africa, and the Chilean-Argentine Peninsula. The Antarctic Consultative Parties have committed themselves to negotiating a minerals regime for the continent. At the Eleventh Antarctic Consultative Parties Treaty Meeting (23 June–7 July, 1981, in Buenos Aires), they adopted Recommendation XI-1, which stated: "A regime on Antarctic mineral resources should be concluded as a matter of urgency." Such a regime "should not prejudice the interests of all mankind in Antarctica," and should "apply to all mineral resource activities taking place on the Antarctic Continent and its adjacent offshore areas. . . ." In 1975 the parties agreed to exercise restraint in mineral-related activities, and there is an unwritten rule that no exploratory or exploitation activity will take place while timely progress is being made toward the adoption of a minerals regime.

It is too early to tell what form an Antarctic minerals regime will take. Some sort of joint management approach seems most likely, but it will be more difficult in this instance to finesse the claimant versus nonclaimant issue. Claimants are likely to demand special consideration based on their claims, which in turn is likely to be rejected by the nonclaimants. One of the advantages of moving to

create a regime to govern mineral exploitation at an early stage is that the location and availability of mineral resources are unknown. By creating a system before states know where deposits are, it is hoped that the claims issues can be handled. At the same time, this lack of knowledge about the scope and economics of Antarctic mining makes it more difficult to formulate a regime that will be viable if and when mining actually takes place.

Once again the real source of momentum on this issue comes from the possibility that the larger international system might intervene. In 1983 the U.N. General Assembly resolved to study the issue of Antarctica. It increases motivation for the Antarctic states to build some type of cooperative approach to the question of mineral resources. As did the Living Marine Resources Agreement, this stretches the 1959 treaty consultative mechanism from its original scientific focus into the broader task of developing suitable resource management schemes. With the United Nations eager to step in, there is a strong incentive to match the success in dealing with living resources with an agreed regime for Antarctic minerals.

Outer Space

In outer space the international community is facing many of the same political and management problems that it has confronted in the oceans and the Antarctic. Outer space includes the use of geostationary and elliptical orbits for activities such as telecommunications, meteorology, resource surveys, or solar power transmission. Initially a few technologically advanced countries had the capability to launch and maintain satellites. In the 1960s the situation was one of relatively few users operating in space, and the legal regime was much like that for navigation on the high seas.

The 1967 Outer Space Treaty was agreed on by the major nations active in space (Gorove, 1973).[2] Few developing countries took part in negotiating this treaty. It adopts the laissez-faire approach, allowing free exploration and use of space subject to

---

[2] Treaty on Principles Governing the Activities of States in the Exploration and Use of Outer Space including the Moon and Other Celestial Bodies.

the proviso that space activities be carried out for the benefit and in the interest of all countries in accordance with international law. Activities are to be pursued on the basis of equality and international cooperation if scientific research is called for. Weapons of mass destruction are proscribed, as are military installations in space. The treaty also bans national appropriation.

Although space still remains an uncrowded and underexploited area, a number of developing countries became apprehensive in the 1970s that certain space resources were, in fact, finite and that the developed countries would appropriate available resources operating under the first-come-first-served principle. In particular, they were concerned about access to geostationary orbits and the allocation of the radio spectrum.

The geostationary orbit is located 22,300 miles above the equator. At this altitude satellites can remain in a fixed position above the earth to carry out a number of telecommunications and related activities. The distance by which satellites in geostationary orbit should be separated to avoid interference with one another appears to be a function of technology. Present technology could allow roughly 180 geostationary satellites. However, the use of carefully located antennae, directed beams, and improved stabilizers, as well as other new technologies, could allow as many as 1,800 satellites to be stationed above the equator. Clearly, the possibility of accommodating 1,800 satellites would pose far fewer allocation problems than the prospect of only 180 (American Society of International Law, 1977).

Although there has been no internationally agreed definition of where national airspace ends and outer space begins, it has been widely accepted that earth-orbiting satellites move in outer space. The lowest orbit for maintaining a satellite is a 62-to-28-mile ellipsis. By this definition, the geostationary orbit would seem to fall well beyond national airspace. Despite the generally accepted norm for defining outer space, eight (of eleven) equatorial nations agreed in the Bogotá Declaration of 1976 to claim the geostationary orbit as part of their national territory. They argued that, in the absence of an explicit definition of outer space, the

ban on national appropriation does not apply to them, even though their claims extend to 22,300 miles above the earth. These countries stated that satellites stationed over them must receive their prior authorization. They considered that the geostationary orbit over the oceans was part of the common heritage of mankind and that an appropriate regime should be developed for it. The Latin American claimants further declared that they were willing to negotiate regional agreements with other Latin American nations to provide access to the geostationary orbit.

These claims have been largely ignored in practice. Nonetheless, they provide a striking example of the partition approach carried to the extreme. They also illustrate the anxiety of developing countries that they will be too late to benefit from common property resources because the technological capabilities to use them are in the hands of a few developed countries.

Similar anxieties are apparent in negotiations over the allocation of the radio spectrum at the World Administrative Radio Conference (WARC). The developing countries prefer to reserve space for themselves even though they are presently unable to use it. The developed nations argue that that is a waste of the spectrum. Instead, they point to the fact that technological progress has continued to allow an increase in the number of wavelengths available to all users. At the 1981 and 1983 WARC negotiations, the developing country perspective prevailed, and certain portions of the spectrum were not assigned in two of the three regions.

Environment

Environmental issues are a unique form of global commons problem. Where a particular body of water (river, lake, ocean) is shared by multiple national users, problems of management arise. Similarly, domestic activities affecting the atmosphere and the stratosphere have impacts on climate, weather, health, and resources of other countries (CEQ and Department of State, 1980).

A variety of ostensibly domestic national activities may adversely affect the atmosphere or the stratosphere. The conversion of forest lands to agriculture, combined with heavy reliance on fossil fuels, can increase levels of carbon dioxide production and the reflectivity of the land. On a substantial scale, such changes may affect the earth's temperature by several degrees. A longer-term concern about carbon dioxide particulate matter is that it will cause a "greenhouse effect" by trapping the sun's energy. National policies on the use of chlorofluorocarbons should also be a matter of international concern if those chemicals do reduce the capacity of the ozone in the stratosphere to protect us from the sun's ultraviolet radiation.

The environmental problems in the atmosphere and stratosphere may be seen as twofold. The easiest to recognize are the relatively localized problems caused by direct impacts of pollutants and particulate matter. The Nordic countries experience acid rain originating from the factories of the United Kingdom, while the United States and Canada contribute acid rain to one another from their industrial areas. We have recently learned that the Arctic haze over Point Barrow, Alaska, comes from the factories of western Europe. Although these cause-effect relationships are relatively easy to determine, there are significant political and economic reasons for avoiding the scientific findings, especially when reducing the impact of pollution on one's neighbors would create significant costs.

It is even more difficult to secure international agreement on the long-range atmospheric or stratospheric impacts of national activities. Verification of a trend toward the warming of the earth would probably not induce governments to modify their policies on deforestation, particularly where these are economically rewarding. The National Academy of Sciences' analysis of the impact of chlorofluorocarbons on the ozone layer was not accepted initially by our European scientific colleagues. Then, during the Reagan administration, the United States adopted the position that we do not yet know enough about the issue. The implications of these findings would clearly require changes in a number of national policies on the use of aerosols, nitrogen fertilizers, and refrigerants and would have economic costs. Before international agreement to act on long-term environmental trends can be achieved, the scientific community will have to es-

tablish a sufficiently compelling agreed basis of facts. The nondivisibility of the waters or the atmosphere drives nations reluctantly in the direction of joint management. The problems must be serious, however, and the scientific and technical findings must be compelling, for policymakers to adopt economically costly environmental protection measures. An important danger is that the policy process may be too slow; it may react only after an environmental trend has become irreversible.

Governments can be expected to move slowly toward joint management in many environmental areas. We can see the beginnings of joint strategies in present bilateral efforts. First damage must be demonstrated and verified; then liability and regulatory schemes must be considered; and the national governments must get involved. Somehow this process also has to be reconciled with national interests. Ultimately, for the broader global approach, fact-finding and monitoring will be absolutely critical to actually getting something done. But documentation will have to be followed by well-timed political initiatives.

In summary, a variety of management approaches have been adopted to deal with the global commons of the oceans, the Antarctic, outer space, and the environment. Laissez-faire, with the resulting overexploitation of the resource, is still the politically easiest approach. However, it is becoming less and less acceptable to many nations. Partition has been and will continue to be used where it is politically, geographically, and economically feasible and where the consequences of national activities can be limited to national areas. Where these conditions do not apply, however, nations will be forced by physical realities to adopt some form of joint management approach.

Discussion

QUESTION. I know what commons are in a general way, but could you be a little more specific about the attributes of a common resource?

ANSWER. Here is the essence of the commons problem as it was seen originally. You own a certain number of cows, which you graze on the city commons. You get a certain amount of meat from them. The more cows you own, the more meat you get. Others can graze the commons as well. The total amount of meat that the commons produces is, of course, a function of your production and everyone else's. The city's problem looks different from yours. As everyone sets out more and more cows to graze, there is more meat up to a certain point. After that point the cows become increasingly slender and production tends to level off. If you keep forcing more cows on the commons, you ruin it entirely and nobody can pasture any cows there. From a certain point on, the total amount of meat drops off as the number of cows goes up. The essence of the problem is that I, being a selfish person, say "I'll just add one more cow because it will more than make up for my share of the group's loss." I expect to get a substantially larger amount of meat from adding that cow. However, everyone else is thinking the same way, and the total amount of meat—mine included—drops.

The only way out of a situation like this is to reach agreement beforehand on the total number of cows the commons can accept and to be sure that no one adds any more. That way, the group as a whole is better off. All the individuals are doing as well as possible without hurting anyone else. This approach is nothing new. The Cambridge, Massachusetts, common, starting in the early 1630s, was used to keep the city's cows. In 1635 a Mr. Cook was given permission by the city fathers to cut some trees down and to keep cows. Because this was commonly held land, if anyone wanted to cut down a tree he had to get permission. The added complication at the international level is that there are no city fathers to settle disputes. The herdsmen have to work out agreements among themselves.

QUESTION. It is hard to find analogies between the global commons and national and regional ones. In the Cambridge case there was a central authority, a central law. Joint management approaches demand a type of authority that just does not exist at the international level. Some states act like spoiled children, and they are likely to continue to act this way for a long while. Although approaches like joint management appeal to me very much at an intellectual level, I do not have much hope for them. There is always national gain to be had from taking advantage of such arrangements. In the few cases where joint management has been tried,

states do take advantage of it. Without a central authority you are lost. Do you see evidence for more optimism in this assessment?

ANSWER. When you look at international group politics, you can see the states as spoiled children; on the other hand, you could look at them as purposeful actors. They are entities that have been set up to represent the interests of their bodies of people, their citizens. They behave in the people's perceived interest. We may say that they are like spoiled children, but there is a clear logic to their actions. Everyone wants to exploit a resource up to the point of diminishing marginal return. Then they want someone else to cut back. In the absence of a central authority, the only hope is a negotiated outcome. To have a negotiation you need basic agreement on the situation. That is where the technical experts are useful. You need agreement that each additional unit of pollution is affecting everyone, for example.

It is indeed more difficult to operate this way, and the role of information becomes all the greater. You need some basis from which to negotiate a mutually beneficial solution to these global environmental problems. Think of it as a challenge.

QUESTION. I am surprised that people cannot see the clear connection between protecting the natural environment and human needs. For example, if the last herring are taken from the North Sea, both the fishing industry and the people who depend on the fish for food are in trouble. Why do you think this is not self-evident?

ANSWER. Although the connection you mention is real, the details of most situations make the one-to-one correspondence between people and resources less exact. For instance, I am worried about the herring as an endangered species. There *are* herring in the North Sea, just not enough of them to make it profitable for anyone to fish there. These fish are an enormously resilient stock. All the other major commercial fish are the same. Still, I am worried about the fishermen too. There are not enough fish to make catching them profitable. I am worried about the people who need fish to eat, and for the same reason: There are not enough herring in the North Sea to make it commercially interesting to provide them to consumers.

In the absence of some unforeseen happening, the herring will be back at commercially attractive levels in five or six generations, about twenty or twenty-five years. Of course, there are second-order effects—there may be changes in species that feed on herring eggs, for example. But the herring themselves will be back. Twenty-five years is nothing for the herring, but think of all the towns in New England that would essentially disappear if there were no fishing industry for twenty-five years. The same will be true in Europe. The human impact is greater than the biological one, and it usually happens sooner. You must examine the details of each case carefully.

You also have to remember that people's perception of what is "best" can be rather narrow. We lack better governmental control of the fishing industry in the United States because the fishermen have marvelous public relations. The national press corps pays close attention to their attitudes, and legislators pay close attention to the press. Unfortunately, it seems that the attitudes of the fishermen are based on how much immediate profit they can make. Their position is not based on considerations of maintaining the stock, or even on the needs of their own children. The fishermen are their own worst enemies. And this is not an isolated case. You face the same problems all the time trying to manage common resources.

QUESTION. You have talked about national and international solutions to some of these problems. I assume you do not hold much hope for regional solutions. Take the example of the Malacca Straits. I wonder if there has been any interest in setting up a regional strategy treaty for managing oil spills and shipping? Could you address both that particular issue and the broader question of regional action?

ANSWER. A regional solution is desirable for regional problems. An area such as an enclosed or semienclosed sea is clearly a place to adopt a regional approach. The Malacca Straits, however, are used by a larger number of states than those in the region. I do not think that such international straits, which are used by countries throughout the world, ought to be dealt with on a regional basis. Indonesia has a direct interest and clearly should be involved in any activity that will affect those straits, but Japan has a vital interest as well although it is far from the region.

There is a responsibility that goes with being adjacent to major straits. You can argue that Spain and Morocco are bearing the costs for everyone's use of Gibraltar, which is a world shipping way. You might be able to work out some means of compensation. The argument you would have to make is that this is the new regime and that the old idea that these are simply international waterways open to anyone does not apply under these changed conditions. Further, you would have to argue that the new regime gives the littoral states—the coastal states—a special interest in the area. Then you would have to determine what costs they are bearing, what responsibilities they are undertaking, and how they can be compensated for these in a way that does not interfere with international use.

QUESTION. The developing countries are saying that they must have industrialization whatever the cost. The potential for environmental problems in Africa is much worse than it is in the developed countries. Is there justification for their attitude? How do you convince countries to take their share of responsibility, especially when they face such important human problems?

ANSWER. In Africa, for example, there are some very poor countries. One must be sympathetic to the difficulties they have providing for the basic needs of their people. The situation is different in the NICs—newly industrialized countries. States such as Brazil are relatively wealthy. Problems are caused by environmentally unsound government policy, not just poor people who are slashing and burning. They have concerted government policies to develop no matter what the cost. The tropics hold genetic stocks that are clearly common resources. Many of their endangered species may be useful for producing new food sources, protecting existing crops from disease, or developing new medicines. These resources are disappearing without any major attempt to save them. Developing countries often have prodevelopment attitudes, and they certainly do not like to be told by developed countries that their policies are not appropriate. We are going to have to begin looking at who gains and who loses if we protect these resources—and begin paying our share to conserve them.

QUESTION. Most of the environmental problems you have mentioned are caused by changes in technology. There is a common supposition that an equal technology will develop to cure them. I also hear you saying that some of these new technologies are going to create even greater disparities between the states that have them and those that do not. I suspect the sharing of technology that will be necessary is going to create major international political problems. Do you agree?

ANSWER. Your point is well taken. I certainly do agree that the political problems will be at least as difficult to solve as the technical ones. You do not even have to look to the international situation to see how these problems occur. Countries such as the United States are investing heavily in the development of pollution control technology, but even U.S. polluters are reluctant to adopt these expensive new technologies.

Returning directly to your question, the international political problems may be extremely hard to solve. How much do you suppose that the United States is willing to trade for a geostationary orbit? That sum would be a direct measure of our unwillingness to help other countries develop satellite programs through technological transfers or direct grants. There are only so many orbits, and we want as many of them as possible.

The traditional economic argument is that scarcity encourages new technology, which creates substitutes for resources that become too costly. But the time lag between the development of the problem and the discovery of the solution (if one *is* found) is a period of stress. Also, the substitute is usually more expensive than anyone would like; and, of course, the solution is the property of the country that develops it. It will be up to the more advanced countries to be sure that the benefits of these new technologies are shared sufficiently, because the other side of the problem is the political instability that results from widening the gap between rich and poor nations.

Tropical Forests and Genetic Resource Areas

THOMAS E. LOVEJOY

ALEXANDER R. BRASH

The world's remaining tropical forests are a most precious asset; their perceived value to society may only be surpassed by the wealth inherent in their myriad of species still unknown to science. Although it is hard to be precise, especially when barely 1.5 million of the world's estimated 8 to 10 million species have been named, approximately 50% of all life forms live in the earth's tropical forests. Each of these species has evolved to fill its own particular niche by adapting to a host of problems encountered in its environment. Every extant species can thus be viewed as having responded successfully to a series of particular problems, and these responses constitute a major reference library to which society can turn for solutions to the wide array of biological problems that continually confront us.

The Ecology of Tropical Forests

The high level of biological diversity (measured as numbers of species) found throughout tropical forests is evident even on a local level. For example, whereas forests of the north temperate zone might contain 10, or perhaps at most 20, species of trees in a plot of about 14 acres (5.5 hectares), a similar-size plot near the mouth of the Amazon, at Belém, Brazil, has been recorded to have 295 tree species (Hatheway, 1971). One immediate consequence of this high diversity is that most species occur in low densities and are widely, and apparently irregularly, dispersed throughout their range.[1]

The pollination systems of tropical forests' trees reflect this type of distribution. Wind is the prevailing method by which temperate trees disperse their pollen. In a tropical forest, however, most pollen would never reach a receptive flower this way; consequently, animal pollinators, a more targeted means, constitute the dominant mode. Usually a coevolutionary arrangement between the pollinator and the pollinated has evolved to their mutual advantage. In exchange for food resources gathered in the course of pollination, the pollinator provides the tree species not only with the transportation requisite for pollination, but with outcrossing, so that genetic heterogeneity is favored.

One such relationship involves the Brazil nut family (Lecythidaceae). The flowers of some less florally advanced genera of the Lecythidaceae are general in construction—they permit a variety of bee species to enter and carry out pollination. Other members of the family (such as the genus *Couratari*) are so specialized that their flowers' intricate configuration permits only one bee species to enter and pollinate. *Couratari atrovinosa*, for example, is believed to be pollinated by a single species, *Eulaema meriana*, a large Euglossine bee. That bee, however, must depend on a number of other tree species for subsistence during the major portion of the year, when the *Couratari* tree is not in flower, thus illustrating the interdependence prevalent in tropical ecosystems. A final complexity in this unique relationship is that it involves only the female bees, for the males have established their own relationship with certain orchids.

There are numerous examples of complex relationships that range far beyond pollination systems. In return for a colony site and often a food source, certain ant species provide a plant's defenses, not only against other animals—particularly herbivorous insects—but also against other plants—for instance, the strangler fig—which are competing with it for light or space on the forest floor (Janzen, 1966, 1969). In certain instances the detritus that accumulates from ant foraging activities provides additional nutrients to the plant and, on occasion, a root substrate. Such relationships between ants and plants occur in at least sixteen neotropical ant families, including thirty-five genera and over 200 species. In Asian tropical forests at least 150 ant species are known to be involved in such relationships, while in Africa so far only 40 or so species have been recorded (Benson, in press).

The prevalence of such complex relationships is not surprising given that tropical forests and their associated fauna date back several tens of millions of years, thus providing plenty of time for intricate interrelationships to develop. Not only bees but some bird and bat species disperse pollen in return for food. Recently rice rats and deer mice (*Oryzomys devius* and *Peromyscus mexicanus*) were discovered to be involved in a similar arrangement with a member of the Melastomataceae family in a Costa Rican forest (Cecile Lumer, personal communication). Many animals, such as bats, birds, and monkeys, and forest floor rodents as well, act as seed dispersal agents for food rewards, but in this case the incentive they receive is the fruit's flesh. In the Amazon many fish species disperse seeds after eating fallen fruits, although some destroy the seeds in the course of devouring them (Goulding, 1980). Sloths appear to supply nutrients for certain trees in their home range by simply concentrating their defecation near the trees' bases (Montgomery and Sunquist, 1975).

Such interdependencies occur not only on the level of a cluster of species, but also on a regional and perhaps even global scale. It is now known that in the Amazon basin, where the world's largest rain forest occurs, water is recycled within the basin to such an extent that about half of the rainfall is generated internally. Most of it originates from the forest itself, rather than coming from the ocean (Salati et al., 1979; Salati and Matsui, 1981). When a forest is cleared, the rain falls directly on the ground, rather than having its descent broken and dispersed by the forest so it will arrive at ground level more gradually. Water also infiltrates the soil considerably more slowly in a cleared area than under forest conditions (Schubart, 1977). The result is runoff, attendant heavy erosion, and a reduction in water entering the soil, which under normal

---

[1] Recent studies (Hubbell, 1979) indicate more clumping within species of tropical forest trees than had previously been thought.

FIG. 12.1. *Three-toed sloths, which provide nutrients for the trees of their home range in Western Hemisphere tropical forests, near Manaus, Brazil. Photograph by R. O. Bierregaard.*

conditions would be returned to the atmosphere by evapotranspiration (through the plants).

This inevitably means that forest clearing beyond a certain threshold will not only reduce water recycling in the basin, but is also likely to trigger a drying trend that will reduce recycling further (Salati, Vose, and Lovejoy, in press). In addition, the probable resultant reduction in cloud formation is likely to affect the global transport of heat and climatic patterns as well, because a major factor affecting the earth's ocean currents and global weather patterns is heat rising into the atmosphere at the equator. Whether similar patterns of water recycling occur in other tropical forest regions is not known. Central American tropical forest regions receive rain that is probably mostly oceanic in origin, but in Southeast Asia and the Zaire (Congo) river basin some degree of water recycling is a possibility. The problems of runoff on nonforested soils are certainly ubiquitous.

Destruction of the Forests

How do people interact with such biological profusion? The classic view of the tropical forest and its ability to recover from the onslaughts of civilization is: Just turn your back on it, and it will take over. However, this is not quite the way it actually works. The forest's fragility has certainly been demonstrated since the beginning of the 1970s, when Brazil decided that it really wanted to use the perceived high biological productivity of the Amazon to support some of its poverty-stricken people. It also wanted to establish itself militarily so that nobody could challenge its authority, so it began the great Trans-Amazon Highway building campaign. The Trans-Amazon is not just a single highway; it is a whole network of roads. Though it never reached as far as was originally planned, one can now drive (with some difficulty) from Manaus, Brazil, to Caracas, Venezuela. New skyscrapers have begun to go up in the major Amazon cities of Belém and Manaus, which was the old rubber boom capital a thousand miles upriver. There is no question that people and development are coming to Amazonia.

FIG. 12.2. *Burning of the previously cut forest near the Minimum Critical Size of Ecosystems Project in Manaus, Brazil. Photograph by R. O. Bierregaard.*

Areas that were largely unbroken forest only ten years ago are being cleared. In many cases the timber is not even being used, because transportation costs make it too expensive to get to market. Instead, the forests are being burned. This almost immediately releases the entire nutrient load that they contain. Because most of the nutrients in the Amazon system are in the living matter, crops and pasture grasses will grow in the first couple of years after a burn, but succeeding years' pastures will deteriorate as a result of the increased invasion of weeds and the continual loss of nutrients (Hecht, 1979; Fearnside, 1979; Lovejoy and Schubart, 1980). By and large, the conversion of tropical forests to pastureland is an unsuccessful operation in Amazonia. Today there is far more abandoned grassland in the state of Pará in the lower Amazon than there is active pasture.

The large-scale burning is clearly adding a fair amount of carbon to the atmosphere, at least temporarily, as well. We do not know how much the deforestation in the tropics is adding to the amount of carbon dioxide in the atmosphere, which is 15% greater today than at the turn of the century. In fact, there is a great deal that we do not know about the dynamics of the world carbon cycle, especially about where and how all the carbon is ultimately assimilated. Yet it is very clear, simply from the increases of carbon in the atmosphere, that the basic systems for maintaining relatively constant carbon dioxide concentrations are unable to cope with the carbon load we are now producing. Considering that tropical forests hold something on the order of 370 billion metric tons of carbon (roughly half the amount of all the carbon in the atmosphere), their burning is indeed something to worry about.

Protection of
Genetic Resources

If one is concerned both about using the tropical forests for satisfying human needs and about protecting the wealth of species, two problems arise immediately: where the protected areas should be and how to conserve species that have never even been observed. A large portion of the flora and much of the invertebrate fauna of the Amazon basin is undescribed.

At present there is a far from adequate short cut, but it is the best available approach for the moment: Biologists look at the distribution patterns of groups of organisms that *are* well known—primarily birds and butterflies. Endemic species (those that occur only in limited areas) within these groups cluster in fairly well-defined spots. It is thought that these clusters define the relict areas where tropical forests persisted during the cold, dry periods of the Pleistocene epoch, when the Amazon forest was not able to maintain itself across the entire basin. It is also thought that these isolated Pleistocene refugia could have been the centers for the evolution of new and different species, and that present endemic species probably evolved in these refuges and have not yet spread out across the rest of the basin for some reason (Haffer, 1969).

Whatever the origin of these clusters of endemics, their areas are important for conservation (Lovejoy, 1982). Relatively few clusters of endemic birds would need protection, but for butterflies the situation is more complicated. A greater number of parks would be necessary to protect the many concentrations of Lepidopteran species with limited distributions. It is important to point out, however, that the entire region need not be a national park to protect the full variety of the species that occur within it. There is room for development or modification of the forest in some areas along with a system of reserves to protect the wealth of plant and animal life in others.

Problems arise, however, when a park or a reserve is established in what once was continuous forest. The newly created reserve becomes an "island" in a "sea" of development. A fragment of forest or any other kind of ecosystem will lose species in an ecosystem decay process (Lovejoy and Oren, 1981) after being isolated. Until now, no one has had the chance to study this process of species loss and ecosystem decay from the moment a fragment of the forest is isolated from the rest. The World Wildlife Fund and INPA, Brazil's National Institute of Amazon Research, have been very fortunate in having an opportunity to do just this 80 kilometers north of Manaus (Lovejoy, 1980). There, a series of predetermined areas of forest are being set aside. While the forest is still continuous, scientists census selected species and groups of plants and animals and then follow their history through and after isolation.

At the outset isolated forest areas receive an influx of species from the surrounding cleared forest as it is destroyed; some mobile species cluster in remaining small intact fragments. They gather in overcrowded concentrations, trapped in a small part of their former range. Two bearded saki monkeys (*Chiropotes satanas*) were thus isolated in a 10-hectare patch. These are animals that normally travel in bands of twenty or more, covering maybe 600 hectares per year. The two isolated ones kept returning to the same tree two or three times a day, eating unripe green fruit. Clearly they were under stress, not a very pleasant experience to observe, but an integral part of the ecosystem decay process under study.

Within just a few weeks the direct effects of isolation, and the presence of new species, change the dynamics inside the smaller forest fragments. Through colonization (with perhaps some recolonization) and considerable local extinction, the decay process leads to an ecosystem that is less diverse and probably more stable in composition than the original. This process will occur rapidly over a few months in a small patch (such as 1 hectare) and quite slowly in a large block (about 10,000 hectares).

The decay process and the related strong spatial component of tropical forest ecology cannot be ignored when new reserves are planned. For example, army ants (*Eciton burchelli*) live in colonies of 500,000 to 1 million individuals, which go through the forest looking under every leaf and running up small tree trunks in search of insects and other ani-

FIG. 12.3. *The one hectare (on left) and 10 hectare isolated reserves of the* MCS *Project in Manaus, Brazil. Photograph by R. O. Bierregaard.*

mal prey. Their hunt is so impressive that pioneering ecologist Charles Elton termed it a most macabre ecological event. A large number of bird species take advantage of the disruption the ants cause. They follow behind a swarm, darting down to pick up insect prey that attempt to flee before the ants get them. These prey species normally sit very still and are well camouflaged, but when forced to flee from the ants, they become obvious targets for the birds.

There is certainly a minimum area of forest necessary to support a given number of army ant colonies, so that a minimum number of them are in swarm on any given day and the birds that depend on the ants' activities can feed on a daily basis. Without the ants, the obligate ant-following birds would soon leave or die. Similarly, the complex arrangements of many tree species and their pollinators depend on a minimum area. Simply to set aside a random-sized tract may overlook many area-de-

pendent factors, leading to loss of many of the area's species.

The Need for Forest Conservation

Amazonia is more than the forest. It is also the river itself, the enormous quantity of water (20% of all the world's river water), and its fishery, estimated to have a standing stock of roughly a million metric tons. This figure can be put into perspective by remembering that the world fishery, taken at its peak in 1970, was something on the order of 70 million metric tons. Studies supported by the World Wildlife Fund since 1980 have documented how many of the major food fish of the Amazon swim into the flooded forests during the high-water months of the year to feed on fruits, seeds, and other living matter that fall into the water (Goulding, 1980). This activity enables these rivers, which are basically nutrient poor or photosynthetically limited by turbidity, to support a far greater fishery than they otherwise might. Some other fish, such as the 2-meter-long giant Amazon catfish, which cannot swim into forests, feed on food chains originating in the forest. Many of the fish that swim into the forests to feed on fruits and seeds eat nothing else during the rest of the year; instead, as the river subsides and returns to the main channels, they subsist on fatty deposits accumulated during the high-water feeding period. Certain piranhas actually join in the seed eating during high-water months.

The major source of animal protein for people living in the Amazon basin comes from these fisheries. They enable Manaus to be one of the few tropical cities in the world that does not have an animal protein deficiency. But the fisheries are threatened. The areas of the floodplain where the forests are, and where the fish go during high-water months, are some of the few sites with any promise for expanding Amazonian agriculture. They get an annual deposit of silt every year that makes them quite fertile. Consequently, they are being considered for rice cultivation. This would involve a trade-off of fish for rice that is attractive at first blush, but would inevitably impoverish Amazonian waters. The World Wildlife Fund is now working with the Brazilian government to gain some measure of protection both for the fishery and the floodplain; there should be ways to utilize floodplain productivity and keep the fishery too.

Returning to the genetic bank consideration and the value of maintaining biological diversity, two species from the Amazon basin have made an enormous impact on civilization. The obvious one is the rubber tree (*Hevea brasiliensis*). Nearly everyone has used one of the 50,000 products from the rubber tree or something created in the laboratory that was inspired by rubber. At one time all the rubber in the world came from the Amazon basin. Because the Amazonians had a monopoly, they could afford everything they wanted: two opera houses, for example. Even Sarah Bernhardt played Manaus.

The second of the two species that has made a major impact on civilization is part of a lesser known story. It involves the giant Amazon water lily (*Victoria amazonica*), which has pads up to 2 meters across supported by elaborate ribbing underneath. The first person to get the giant water lily to bloom in Europe, in the early part of the nineteenth century, was Joseph Paxton, head gardener for the duke of Devonshire. He was enormously impressed by these water lilies and all their structure, and he felt they must be capable of supporting a great deal of weight. After he experimented and found that one of their pads could support at least 60 pounds, he decided to build a small greenhouse following the structural principles derived from the bottom of the lily pad. Paxton later used the same principles to design the Crystal Palace of the great mid-nineteenth-century London Exhibition—the building that is acknowledged as the beginning of modern metal-beam architecture.

The capacity of tropical forests to inspire new invention or provide new resources is far from exhausted. It has hardly been scratched. Recently, carrying a cupuacu (*Theobroma grandiflora*), a relative of the cacao, back on an airplane from Brazil, I touched off an experience like being in the Land of Oz. Everyone was wandering around the cabin, sniffing, and wondering what the marvelous scent could be. This wonderful fruit has never been widely used commercially because it is filled with heavy seeds and large tendrils that make it difficult to process. Within the last two years, however, scientists from INPA have discovered a seedless variety of the cupuacu in the forest. The director of the INPA likes to make the point that we could have spent millions of dollars trying to get a seedless variety, and here the Amazon forests had it done for free.

Discussion

QUESTION. You have calculated that as many as 20% of all species could be lost by the year 2000. Could you tell us, in general terms, how you arrived at this alarming figure? I do not doubt that it is a good estimate, but it is still a staggering total.

ANSWER. The rate of species extinctions is tied directly to the rate of habitat destruction. To understand why so many species will be lost, it is necessary to look at what is happening to the places where they live.

At present, something on the order of 10% of the tropical forests of the Amazon have been cut. By the year 2000, under a reasonable set of assumptions, at least 50% will be gone, probably more. Some estimates come very close to 100% loss. Very few of the species that live in these forests will be able to persist if they are driven beyond their current range limits; they will simply die out. In other parts of the tropics the situation is not much different. There will be vast areas deforested in Africa and southeast Asia. Outside the tropics the rate of habitat destruction is likely to be lower, but still considerable.

The number of species lost is going to be a function of the amount of habitat lost. We can make different assumptions about the shape of the curve that describes this function, but it is likely to be nearly linear and its slope is going to be nearly one—for a 50% loss of habitat the species loss in that area will also be somewhere around 50%.

Totaling the results of all these depressing calculations, we project that 15% to 20% of all the species on earth will be gone in another fifteen years or so. This is the same as saying that there will be 500,000–600,000 extinctions before our children become adults. Even if this figure is wrong by 10% or 20% of the total, it is still appalling.

QUESTION. Slash and burn agriculture seems to be the small-scale equivalent of clearing tropical forests for pasture. Is it as damaging to small areas as large-scale deforestation is to the forest as a whole? Can the poorer tropical soils be used for any agriculture at all?

ANSWER. Slash and burn agriculture is not inherently a bad idea. If you look at the way the forest recovers after one period of cultivation, you find that the processes at work are not very different from those that operate as the forest recovers from a small fire or other natural disturbance. Some nutrients released from the native vegetation are taken up by crops, such as manioc (cassava), which is one of the major harvests. The rest of the nutrients are lost to runoff and leaching. But if the area is not disturbed again, the natural forest will return and the nutrient cycle will recover. This

should not be surprising. There is always evidence of recent disruptions in tropical forests, but the signs of older ones are hard to find. At any given time some parts of the Amazon basin are recovering from minor natural disturbances.

The real problem comes when the population subsisting on a slash and burn agricultural base becomes so large that the forest does not have time to recover fully before the next cycle of cultivation begins. Once the population passes the critical point, at which the runoff and leaching I referred to become constant, the whole nutrient budget is thrown into disarray. The forest—both vegetation and soils—degrades rapidly. It would be very interesting to know just how much disturbance can occur before this process of decline is no longer reversed by natural mechanisms. But it is probably more important to find ways of providing sustainable sources of food for the increasing number of the poor who are putting such pressure on the forest in the first place.

Large-scale deforestation is a different matter entirely. If the area is used for pasturing cattle, for example, not only are there the same problems with nutrient loss, but the soil structure almost immediately becomes more compact. When the rains come—and they always do—there is much more runoff over the surface because the water cannot be absorbed by the denser soils. This in turn leads to an increase in the rate of soil erosion. These problems, together with the weeds that invariably invade open pasture, keep large-scale husbandry from becoming a sustainable enterprise. All too often, land is cleared for pasture and then abandoned in a few years.

QUESTION. What can be said about the condition of parts of the tropics where the soils are better or the rains less severe? Tropical highland areas are probably in greater threat than the lowland basins. Because these areas can be used easily to grow market crops, it seems that many ministries of agriculture are anxious to see them stripped of their native vegetation as quickly as possible.

ANSWER. Some of these areas can be very productive, but it is dangerous to rush into converting them into farmland as fast as is being done now. The example I use most often to demonstrate this concerns a perennial species of wild corn that was discovered in the mountains of Jalisco, Mexico, just a few years ago. If its potential for revolutionizing corn agriculture is ever realized, it will be a tremendous boon to society. First, think of the enormous savings we would have in time and money if there were "corn orchards" rather than corn fields. It would not be necessary to till and plant every year, less fertilizer would be needed, and soil erosion problems would be cut back dramatically. Even under present conditions, corn is the third most important grain crop in the world. How important would it become if it were a perennial crop? Not only is the life cycle of this species, as it is encoded on its DNA, important to us, but its chromosomes also carry resistance to four of the major diseases that attack cultivated corn. If we can breed those resistances into annual corn that will be a major accomplishment in itself, one with far-reaching implications for human well-being.

I would argue that the world would be a much poorer place if that hillside had been converted to an ordinary corn patch before anyone had the chance to investigate the wild plants growing there. We do not live on a well-known planet. There are undoubtedly cures for human diseases, new foods, and probably whole new industries whose secrets have not yet been discovered. It is a terrible mistake to let the last of a species that holds such riches be lost for the sake of short-term gain.

QUESTION. Aren't North Americans and Europeans being hypocritical when we tell people in the Amazon not to develop? Changing the landscape radically has been successful for us. We have accepted the loss in diversity and other problems that development causes.

ANSWER. When we went to Central America, our first goal was to learn how to farm the way the native people do—and as well as they do. We got them to teach us. When we reached the point that we could duplicate their success, they were willing to listen to us and trusted the new ideas we presented. It is a fundamental idea. If you do it yourself, then you have the right to suggest it to others. You have to prove yourself. It is a rite of passage. One of the dangers with international environmental policy in the Third World is that it is easily misinterpreted as an "us-them" situation, as neoimperialism. But that is not our attitude at all. All the World Wildlife Fund projects in the Amazon are joint studies conducted with Brazilian scientists.

There is also the question of the global significance of tropical forest destruction. We have not lost much biological diversity in the United States, partly because our ecological systems are more resilient. In global terms, the problems in the Amazon are much larger. But what it really comes down to is that no nation acting sensibly should be destroying its biological systems. The Brazilians have just as much right to say to us that we should not be allowing such huge quantities of our agricultural soil to be washed away every year.

Regional Strategies for Managing the Oceans

SIDNEY HOLT

In this chapter I will examine some of the major problems that have arisen during the last three decades of marine conservation history. Your world and mine has changed very dramatically since I first traveled to New England in 1949 aboard the *Queen Mary*. I recently went to Boston again on a Boeing 747. The differences are not only in the technology of transport, or the technology of communications, although these underlie the institutional changes and problems that the World Conservation Strategy addresses. I would like to show you, by analyzing ocean use, that my observation that the world has changed is not trivial.

National, Regional, and Global Perspectives

We can characterize our thoughts as a web woven of three strands. One strand is the national way of thinking. Another is the global way of thinking, and the third, as you might guess from this chapter's title, is the regional way of thinking. I want to demonstrate that the happenings of the last thirty years are an outcome of the interactions among these three ways of thought about our planet. I also want to emphasize that regional strategies, which are the theme of the last section of the World Conservation Strategy, cannot be devised or implemented without taking proper account of global and national ways of thinking about the planet.

A traditional map of the world fragments the oceans to keep the land masses intact. But *my*

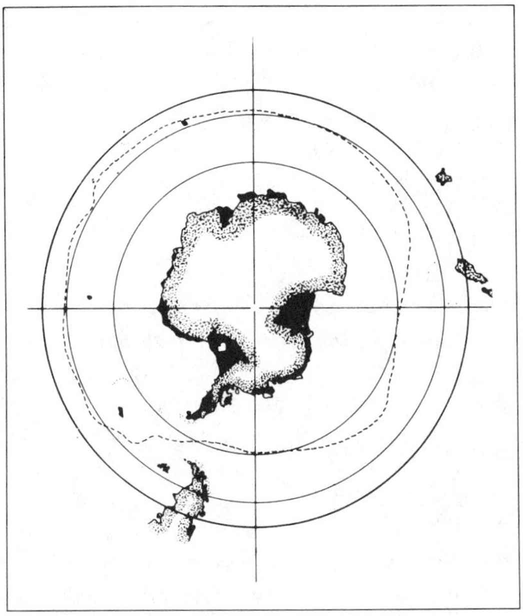

FIG. 13.1. *A whale might view the world as a series of water bodies surrounding the South Pole. Illustration from IUCN, 1980.*

world is divided differently. I do not mind fragmenting all the land masses—almost without exception—as long as I keep the oceans intact. Even this rather peculiar view of the world is a *human* view, however, and one might wonder how an animal such as a sperm whale considers its world. I think it might be inclined to consider the world as a series of water bodies surrounding the South Pole. The Antarctic, which is now so important to us politically and strategically, and may be important economically in the future, would be squarely in the middle of its universe. The sperm whale has been on this planet for tens of millions of years—not only longer than humans, but much longer than most of the primates. Its view of the world has been developing for so long that the locations of the continents have changed quite considerably during that time. So the sperm whale has seen the world undergo changes of enormously greater magnitude than have we humans. My point is that the nature of the world is in the eye of the observer. And our problem, when we start talking about conservation and management, is to decide what our

point of view is going to be. Further, we must recognize that very many other points of view will have to be reconciled, even once we have chosen our own.

A Fisheries Example

I mentioned that I came across the Atlantic in 1949, and I think it is appropriate to say why. I was essentially the messenger of a British delegation to the United States. We were here to negotiate the treaty that set up the International Commission on Northwest Atlantic Fisheries (ICNAF). The purpose of the commission was to try to save the fish stocks of the northwest Atlantic from the depletion that was being noticed on Georges Bank. We had already experienced far greater depletion in the North Sea, and everyone was anxious to avoid another similar episode. At that time U.S. fishing was limited primarily to haddock, the Portuguese were only fishing from small dories, and the Soviet Union was hardly operating at all. There were certainly no factory ships from Poland, Japan, and East Germany working in the area; in fact there were no factory ships.

The Georges Bank haddock had shown some decline, and we were concerned with this in a very narrow sense. No one would have mentioned the word *pollution*. We did not even talk about fishery control, though a few of us wanted to. The British particularly wanted to limit the number of vessels that were operating—we had lost most of our vessels during the war and preferred to stay small while restricting everybody else. It was a typical British position, but, as I learned, no more hypocritical than those of most nations in international discussions. Our only concern was the possibility of making the meshes of trawler nets bigger, so that they would let younger fish out. This would allow the fish to grow, and we could catch them when they were bigger and had produced offspring of their own. After thirty years of discussion and a vast amount of research, the haddock on Georges Bank have almost disappeared—while we have been discussing them. The ICNAF did not have any of the necessary powers to protect the fisheries from overexploitation.

FIG. 13.2. *Fishermen on the Georges Bank. Despite almost thirty years of discussion and research, many of the most desirable species are almost gone. Photograph from Massachusetts Audubon Society.*

In our 1949 discussions we had to cope with a new American invention—the maximum sustainable yield concept. The fundamental idea was that we could regulate the use of a living resource to get, forever and ever, the largest possible catch. Other nations were thinking about solutions to much more immediate problems; this divergence has marked international negotiations for thirty years. The ideology of seeking ultimate goals remains in conflict with the ideology of acting to improve the present situation. In almost every treaty or discussion concerning conservation at the regional level, misunderstandings have arisen from these two attitudes. Perhaps the only difference today is that one

attitude is no longer exclusively North American and the other European. The whole matter is now very much more complex than that. Still, those two approaches are to be found in different weights and with different implications in every one of the 100 or 150 international treaties that now regulate the activities of man in the sea on a regional or global basis.

While we were struggling to form the ICNAF, technology was moving faster than our legal instrument. In more recent years factory ships have come to the North Atlantic from all over the world. Fish stocks have been depleted despite our best talk. All the countries bordering the area have gone beyond ICNAF and claimed 200-mile-wide zones for their exclusive fishing. At the same time offshore oil drilling is about to begin. Our attempts to deal with these fishery problems by looking at one species of fish at a time and trying to conserve them independently have not been successful. We find interactions between species that confound simplistic approaches. This is something general ecologists have been telling us for a very long time, but practitioners have chosen to ignore, mainly because they did not know how to handle the complex interactions in ecosystems. In fact, we still don't know how to handle them or whether we have any possibility of modeling whole ecosystems.

The Global Fisheries Pattern

The ICNAF was one of the first fishery commissions set up (the very first was in the Northeast Atlantic, where a scientific body had existed since the beginning of the century and a regulatory body was established immediately after World War II). But, as usual, the scientists took a broader view than the politicians and the administrators, geographically at least, and were arrogant enough to feel that they did not need to define their region of interest. So in Copenhagen the International Council for the Exploration of the Sea pretended to be what its title implies—an international scientific body. In fact, it only deals with one geographic area of the sea. But at least that area is the whole North Atlantic and not the pieces, east and west, with which we are constrained to deal when we address the political issues of regulation. And this illustrates another generality about all the regional arrangements that have been made in the last thirty years: The scientific view tends to be much broader geographically than the political arrangements. This fact has immense implications in international affairs.

By 1960 the whole of the Northern Hemisphere was covered with fishery regulatory commissions—two in the North Atlantic and several in the North Pacific. These commissions primarily involve the developed countries—what we think of now in political terms as "the rich north." Nearer the equator things began to get complicated. Fishery arrangements came a little later, and a conflict immediately evolved among nations that were unequal in political and economic terms. North of the equator we have the rich Europeans, on the south the much poorer Africans. And just as the continental plates of the earth's surface grind together in the Mediterranean, the north and the south grind together and cause political upheavals. There were attempts to create regional arrangements for the Mediterranean, which for the first time in conservation history involved interactions between the northern and southern nations. Similar attempts were made in the eastern tropical Pacific, where there was a need to conserve tuna resources. The conflict there was between the North and South Americans.

In the Southern Hemisphere many countries are less wealthy, and for this reason the history of fishery controls has been quite different; it has involved more recent and less effective regional cooperation mediated by the United Nations. Now there are regional fishery commissions throughout the whole of the Southern Hemisphere all under U.N. Food and Agriculture Organization supervision or sponsorship.

The regional organizations were established, to a large extent, to help the individual and less developed nations in those areas become competent to deal with their fishery problems. One exception to this pattern was in the southeastern Pacific, where Peru, Chile, and Ecuador established their own commission to control fisheries. They did so because of the Truman Declaration, which in 1947 said that the continental shelves around the United States, or rather the resources on them, be-

long to the United States. The nearly instant response of those Latin American countries was to create their own broad zones of influence in the sea. And thus started a movement that has spread ever since. Since the late 1970s, coastal nations have effectively appropriated large areas of the sea. That appropriation, endorsed in the last years of the Third U.N. Conference on the Law of the Sea, has dramatically and permanently changed the problems of regional cooperation not only for fisheries, but for every other kind of human activity in the sea. We are still watching the consequences unfold.

One large sea region was not covered by a regional fisheries body until very recently—the South Pacific Islands area. It encompasses a group of small, newly independent states which suddenly became big states in the sense that their sea areas are huge under the new law of the sea. Their 200-mile zones all touch one another, so there are thousands of miles of new frontiers. This has led to the proposal by those states to form a completely new kind of regional fishery arrangement, called the South Pacific Agency, which is actually a coalition of those states, to control the whole sea area and ensure, as far as they are able to do so, that the riches from the area will accrue to them. This is the latest stage in an evolution of regional fishery controls. It has led from arrangements among industrial states in the northern Atlantic to the collective occupation of huge sea areas by tiny South Pacific island states. That whole history has taken thirty years, but since the late 1970s the change has accelerated remarkably.

I said this South Pacific area was the last to be dealt with, but that was not quite correct. The sea around the Antarctic, the Southern Ocean, is now being divided. There again, the mainly Northern Hemisphere states, the rich states, that claim Antarctica or claim special rights in it, are trying (and with some success, despite the world's new political nature) to claim rights in the seas around Antarctica. But legally at least the Southern Ocean is still "high seas," open to anyone for exploitation.

The whole of the world's fisheries are dealt with on a regional basis, but these regional arrangements are quite different from one another. It is very important to understand this, because if we looked at the idea of regional strategy, we could very easily conclude that a homogeneous pattern is emerging in the world, when in fact the pattern is extremely complex, and rapidly evolving. Consequently, the World Conservation Strategy is not really going to be able to define *regionalization*.

Whales and Whaling: An Object Lesson

There is only one example in the fishing sector where it turned out that we needed to take a global view. That is whales. The International Whaling Commission was established in 1946, at just about the same time as many of the regional bodies I have mentioned, but it was set up as a worldwide commission. This was because a few Northern Hemisphere countries, having depleted the whales of their Hemisphere, had begun exploiting those of the Southern Hemisphere. The specific political situation demanded a global arrangement. Of course, the other part of the problem is that whales move around a great deal; they are what we now call a "highly migratory" species.

The Whaling Commission has been a highly visible arrangement for fisheries control for many years. One reason for its visibility is that it has been shaken by periodic crises. Every few years countries walk out or there is an outcry about what the commission is or is not doing. Second, the whales became symbols of the whole environmental movement in 1972 at the U.N. Conference on the Human Environment in Stockholm. The third, and a very important, reason for the commission's visibility is an offshoot of what American foreign policy calls the "domino theory." This theory, of course, was originally applied to the feared advance of communism. It assumes that if you "let one country go red" the next one will fall and so on. It is no accident that in regional fishery arrangements, in which the United States has taken a lead from the beginning, the domino theory was applied. The theory says that if the Whaling Commission fails, then the other fishery commissions, the whole structure of arrangements in which the United States has played a very important part, would one by one collapse. (Note that I am not endorsing this analogy.)

The Whaling Commission is visible once again because it is addressing very specific questions that apply to all discussions of regional arrangements (concerning fisheries, pollution, seabed mining, and so on). The first question is, Who can be the members of such a body? Suppose you agree that members must be governments, recognized states. Recently there were arguments about whether Taiwan is a state or not. Then you must ask, What kinds of states are allowed to be members? In some fisheries bodies there have been political exclusions; for example, the Soviet Union is not allowed to join the North Pacific Fishery Commission for purely political reasons. More generally the argument centers on whether or not those who exploit a resource have the right to control its use. The Whaling Commission was established mainly by a class of whaling countries. Since its establishment, some of those countries have stopped whaling. Have they as much right as they once did to determine how many whales are caught each year? Or is this to be decided only by the whalers? And if the ex-whalers have equal rights in that decision, what about the countries that did not bother to join? India and China have recently joined the International Whaling Commission. There are murmurs that this is unreasonable—why should countries that have never had anything to do with whaling be allowed to join the commission and decide how many whales the Japanese and the Soviets can catch each year?

We have exactly the same issue in the Antarctic. There is a fisheries commission for the Southern Ocean whose membership is, under the present treaty, limited to countries that have permanent research stations on the Antarctic continent. Many other countries are asking what this has to do with whether a country should be allowed to fish in the waters around the continent. So membership might be extended to those countries that happen to be fishing there now as well as the ones doing exploratory work. Those who, either technologically or for other reasons, are not operating in that area are thereby excluded. In all our discussions of regional management strategies, we have to look very carefully at our philosophy on who may participate in the implementations. Particularly, we must ask whether to include only nations, only coastal

FIG. 13.3. *The aftermath of a whale hunt. Although humans have traditionally used whales as resources exploitable for their meat and oil, respect for the educational and symbolic value of whales is growing. Photograph by O. Root from Massachusetts Audubon Society.*

nations in that area, only exploiting nations, or only rich countries that can afford to explore.

Common Problems

Practically all the pressing questions about conservation of the marine environment are now being discussed by the International Whaling Commission, although other bodies will have to answer them as well. One of these questions is the role of science in marine management. For the last thirty years the general policy has been that we must base all our decisions on science. Sometimes this has been interpreted as, We must base all our decisions *only* on science. That, for example, has been the United States' position in all fisheries commissions until recently. It has said, in effect, "You must not take account of economic or social practice, at least not explicitly." But we are finding that our science, no matter how much money we spend on research, remains inadequate to the task. We are always confronted with extremely difficult questions about how to act in the face of great uncertainty.

There are further problems about procedures for decision making. Should it be by consensus or by vote? This problem is causing real difficulties in every regional arrangement for marine affairs. In the Whaling Commission we are beginning to have difficulty deciding what is a resource. The traditional view is that whales are there for us to eat—for oil and meat. But some people are noting that we can now make money from whales by taking boatloads of people to watch them and selling T-shirts with their pictures. And others are saying that whales should not be exploited at all, that from an economic point of view they should be treated as nonresources. Now the whales are very special, they are intelligent animals, they are beautiful, they are symbolic, and they are not alone in being objects of the natural world whose value may be as more than exploitable resources.

There are now attempts to find new compromises between those who feel that if a resource is there it must be exploited, and those who feel that maybe it need not be exploited yet, if ever. A particular form of compromise among these ideas is the notion of *sanctuary*. Last year the International Whaling Commission declared the whole Indian Ocean a sanctuary for whales. Some nations have declared zones under their jurisdiction as sanctuaries for whales and dolphins. In its seabed mining legislation last year, the United States was the first country to institute sanctuaries in the deep seabed where mining may not take place. In a very short time we have seen broad application of the idea of sanctuaries, which had been most talked about with respect to whales. Today there are even proposed sanctuaries for manganese nodules on the seabed. I believe this development reflects a search for a stable situation that will last a few years while we get our minds and our economic systems in order to see what our overall view of the planet and our place on it is going to be.

The Whaling Commission has shown us very dramatically that we cannot regulate the whales one species at a time and that we cannot regulate whaling unless we also regulate the capture of the whales' food. More than any other body, it has demonstrated to us, if we needed to have it brought home, the interconnectedness of everything in the marine environment.

The New Law
of the Sea

The same kinds of interplay between global, regional, and national considerations have affected the Law of the Sea. In 1958 the first U.N. Conference on the Law of the Sea was held. It reaffirmed global freedoms of action by those who had the means to be free; it emphasized the continental shelf idea, which the United States had popularized through the Truman Declaration; and it tried—and failed—to fix a narrow limit on territorial seas. As an afterthought, it added a resolution that all this should be put into effect through regional arrangements if possible.

Later conferences went further. The first conference was essentially looking only at the global aspect of the Law of the Sea. This global view was also in many minds when Malta started the procedure that led to the Third Conference on the Law of the Sea in 1969. Malta wanted to see a transition from an open commons to a regulated commons, at least in the deep sea. Its motivation illustrates one of the general laws in marine management: Technology always moves much faster than legal arrangements to regulate it. The technology of deep-sea mining had moved so fast that the 1958 treaties were out of date before they could be ratified. By 1969 deep-sea mining in the deepest water was a clear technological possibility. Very quickly after that the national view took over. The Third World countries looked at the Maltese proposal to declare a "common heritage" for the whole deep ocean and said, "That looks fine, but we would rather extend our jurisdiction to 200 miles now and make sure we get that little bit, because if we have a U.N. organization to run the sea for us, it will be controlled by the powerful countries." Then, very quickly, jurisdictions were extended and a virtual appropriation of the sea occurred worldwide, reaching as it now has done out to 200 miles and sometimes farther.

We have again reached the stage in the Law of the Sea conference at which regional views are becoming important. (At least we *had*, before the Reagan administration threw a spanner in the works). I do not think that the strident U.S. position is going to have a dramatic effect. It will slow negotiations, but in fact the issues are so big—so world-shaking with respect to military strategy and future development—that no national administration, even one as powerful as the United States, can do more than cause a momentary pause in the course of ocean policy.

Initially, there was a *global* view in the Law of the Sea Conference. It showed, incidentally, that very small nations (such as Malta) can cause dramatic policy shifts. Then there was a *national* reaction, in terms of extensions of jurisdiction. But ultimately nations have realized that much of the application of the Law of the Sea is going to have to be on a regional basis.

However, the first effect of these Law of the Sea developments has been to change the *scale* of the regions we think about in the sea. I believe that current developments are bringing us closer to managing the ocean in much smaller pieces. We were used to thinking of big sea areas, like the northwest Atlantic or even the whole north Atlantic. The new Law of the Sea has led to a huge increase in international frontiers—in the number of frontiers, in the number of countries that interface with each other because they now have frontiers at sea, and in the lengths of the boundaries between them. All these frontiers have multiplied many times. As a result, countries are now beginning the process, which I think will continue for at least another decade, of evolving new regional strategies. These involve negotiating regional arrangements, which often concern only two or three adjacent countries—not the big areas we are used to thinking about. This is very much like management of river systems, where two or three countries work together. It is dramatically different from the huge numbers of countries and vast areas that we are accustomed to negotiating with in ocean management.

Multiple Uses of Spaces and Resources

It should be obvious that it is no longer feasible to regulate one use of the sea at a time. I will not describe in detail the other histories that parallel fisheries management, but there have been similar developments in many areas: Since the Stockholm conference in 1972 we have many regional agreements on pollution. Once the new Law of the Sea is in force we will have regional agreements on seabed mining. And we have regional agreements on ocean science. For example, in 1960 UNESCO started its International Oceanographic Commission, which promotes marine science everywhere and conducts cooperative investigations. It was conceived globally, but it now operates mainly regionally. But while regionalism is becoming stronger, the global approach is still in evidence.

For example, the U.N. Environment Program develops regional pollution control organizations, but we still have a U.N. agency that deals with pollution from ships on a global basis.

All our attempts to regulate, to have conservation strategies on a regional basis, are plagued by the complications of simultaneous global and regional approaches to the same problems. Every year it is becoming less practicable to regulate the sea one use at a time. More and more, oil drilling interferes with fisheries; deep seabed mining interferes with maritime transport; and pollution interferes with everything. In the past we could manage one use of the sea at a time, but from this decade onward, we will not be able to. Yet we have no appropriate mechanisms, no legal arrangements for controlling multiple uses in sea areas. We can only hope that some of the experience we have gained trying to regulate the use of water in river basins will be applicable to the sea.

The Scales of Conservation

I wish to offer one last example of a conservation problem whose "scale" is rather difficult to define. The Indian Ocean was historically viewed in a small scale by the Arabs, who were exploiting features of its oceanography and meteorology by moving their trading vessels along the northwestern coasts. A little later, trade expanded to encompass the entire Indian Ocean. This was made possible by greater knowledge of the meteorology and oceanography of the southerly latitudes. Today the entire Indian Ocean (which, as I observed, is a whale sanctuary) is covered by a lattice of trade routes. In fact, the main oil tanker routes pass directly through the areas where Yankee and European whalers found the largest concentrations of whales in the nineteenth century. By putting together reports, maps, and what little we know about the oceanography of the area, we can show that the areas likely to be polluted by tankers are also the areas with the greatest biological productivity.

How will the whales survive? In the Indian Ocean there is, or soon will be, a piece of a global strategy administered by the International Whaling Commission, a regional species protection network, smaller subregional habitat protection areas under the jurisdiction of individual countries, and national multiple-use controls—a nesting of many scales, for a variety of purposes, in a diversity of political, economic, and national situations.

It would be pleasant to offer a view of ocean management in which problems are well defined, boundaries neatly drawn, and responsibilities clearly set. But the present reality is that ocean management is extraordinarily confused. It is not clear to me how the whales will survive or how any other aspect of marine affairs can be controlled as things are today. There is some hope in the trend toward smaller regional strategies in which a few countries come together to manage all the resources within their new common frontiers, but it is too soon to know whether this will do more than add yet one more scale for confusion.

Discussion

QUESTION. Where is the leadership going to come from on questions of ocean policy? At one time America and the European countries seemed to be the real forces behind international action, but recently their efforts are sporadic at best. The law of the sea conference is probably the best example in this regard. I do not see any countries trying hard to make the conference work.

ANSWER. Yes, it is true that the Europeans and the United States have shown leadership on certain issues and in certain areas. But in large measure we have lost that leadership position. The Third World countries are aware of this. If the United States showed the least bit of commitment, many countries would come along with us. In the last few years, I have seen too many places where the United States falters because of its own internal gutlessness. This leaves our traditional allies confused. They do not trust us to stay with any position that is politically risky.

The NGOs that I know from Third World countries have begun to assert leadership in their own nations and to help other developing nations take positive conservation initiatives. The whaling problem is a good example. In many cases you can ask, "What have they got to lose? They can look wonderful. They don't kill whales, they don't even have whales." But when these countries take a very strong stand, they encourage other developing nations to do the same. In some recent instances, their whaling protests have been extremely well documented with help from eminent scientists. They have made sure that their position is both good science and good conservation.

The United States and Europe are being left behind. The representatives of Third World countries are saying, "So what if the developed countries are indifferent to these problems, we can solve them. We have other allies."

QUESTION. Is there any hope for developing regional watershed management strategies that really work? They seem to be a good idea from an ecological perspective, but I think there are going to be major political problems to overcome before they are accepted. Are there good working models for politically acceptable regional strategies?

ANSWER. When I think of a river I think of three different aspects—land along the river, the quality of the water in the river, and the amount of water in the river. We have institutional and legal mechanisms for controlling or regulating each of these aspects. The problem is that they are not coordinated. They are fractured among the federal, state, and local governments and private citizens. There is no comprehensive look at how the entire ecosystem should be managed. Most multipurpose commissions that could have any real effect do not work, because they are afraid to give any of their jurisdiction to new political entities. It is the same at the international level—countries do not yield real power to international organizations.

The Delaware River Commission is at least one example of an institution that has shown some ability to deal effectively with both planning and managing a whole watershed. It was originally set up to allocate water on the Delaware because of a substantial dispute among the states of New York, New Jersey, and Pennsylvania about drinking water supplies. The commission allowed these states to bypass a Supreme Court ruling that would have allocated the water by federal fiat. The governing board is composed of a representative from each of the states and a single federal representative, all of whom have equal votes. It has been quite successful in developing a multiple-use plan for the entire river basin. This includes water pollution control plans, drinking water allocation plans, and management of the water as an energy resource. The commission does not have enormous authority, but it has had some success in convincing the political powers to support its decisions. Its greatest success was probably the implementation of an emergency drought plan. It was able to persuade the state of New York to cut back its water supplies tremendously for the sake of users downstream.

In most cases, though, I would say that the history of interstate commissions in this country has been much more one of failures than of successes. When we are able to convince people to think in terms of whole ecosystems, we will have an easier time persuading them to adopt ecosystem management. Our first job, creating that shift in attitude, will be the hardest.

QUESTION. I worry that integrated management is not the full answer to water resource problems. Can you explain why it will not simply lead to new "governments" with the same priorities as the old ones?

ANSWER. I wish I could explain that to you, but I suspect you are right. We have managed ocean mining, fisheries, river basins, species exploitation, and most other environmental problems one at a time. This is in direct contrast to a more holistic world view. As many times as I have seen photos of that little blue marble from space, I still find them absolutely breathtaking. This planet is where we live. What are we doing with it? We keep missing the holistic view of the earth as an entity because we chop it into tiny, easy-to-manage pieces.

I often worry about regional management schemes, I even worry about global ones, such as the International Whaling Commission. Most of them are based on a Cartesian ethic, a totally utilitarian perspective, which the developed countries try to impose on the developing ones.

I first realized this after a very moving experience and have been thinking about it ever since. It happened at the New England Regional Conference. James Aldridge, editor of the *Environmentalist*, chaired a workshop on international environmental education problems. Someone asked, "How do you explain to people that they should value wildlife?" The very next day someone posed that question again. It occurred to me that in all my discussion about environmental protection with hundreds of people from the Third World, not one of them has asked, "What is it good for?" Not one. I thought to myself, "They are not looking for reasons to value nature, they immediately assume its value." It sounds like a good idea to form international regional commissions, to gather nations together—rich and poor alike—when we decide on major policy issues. But these commissions almost always impose seventeenth-century French elitist philosophy, which says that the world is a marketplace. If we do not work for a more holistic outlook, we are going to lose everything that does not have a price tag attached to it. We will have a world that few of us will want to live in.

QUESTION. I am interested in hearing more about the role of science in ocean policy. I have never thought that science is particularly objective, no matter what people say, because it is conducted by human beings with political and cultural values. Would you agree that these values influence what answers scientists attempt to find?

ANSWER. There is a more basic problem. Scientists have biases based on the fact that they are human beings. The work of the International Whaling Commission is a nearly perfect illustration. The constitution of the commission mandates that it operate from a scientific basis. It did not do this very much in the beginning, because the rules were not laid down very precisely. But in 1974 the commission agreed on very precise rules that flowed from an assumption about what we do with science. The rules said, "You will look at populations of whales, and you will count them, and you will determine their status. If these are your results, you will catch this many, and if those are your results then you protect them," and so on. That protocol led to six years of very intensive scientific work. Theoretically, in science the more you research and the more you argue, the more you converge on the right answer. What has actually happened is that we have diverged and become less and less certain of what we are doing. For a few years the scientists behaved extremely defensively, as do experts in almost any area of government. When they are threatened, they dig in their heels and cover the cracks of their uncertainty. In the last year or two, this uncertainty has become obvious. The more honest among the scientists (and there are some) admit that they had been covering their doubts because they were afraid that a decent assessment of their work would reveal its inadequacies. We still do not really know how to avoid this sort of problem.

One approach I favor strongly is a better but different kind of science. Today we regulate the catch of every species of whale separately. Yet our science, as bad as it is, tells us that the sea is a complex of relations between one species and another. We are at the stage of trying desperately to agree on ways to build new models of what the sea is like.

We also have to change our assumptions about what animals are. We scientists have usually worked on the basis of our own feelings about the world and what affects us and then transferred those feelings to other living organisms, or we have assumed that the animals are not alive at all, that their behavior is totally random. For example, whales probably are affected much more by noise pollution than they are by anything else. We cannot prove it, but we have to think this way. What matters to a whale? Not what it sees, but what it is hearing all the time. So we really need to start looking at subjects such as the effects of oil slicks on a whale's hearing. The interesting thing about whales is that we can actually begin to ask them these things, such as Can you see that oil slick? or Would you avoid it if you had the chance? Their answers are fascinating. They say that not only can they see an oil slick as well as we can, but they can hear it. The little pieces of data begin to make us think very carefully about our initial assumptions and experimental designs. How should we react to the knowledge that whales can use their sonar to determine that there is a one-molecule-thick film of oil on the ocean surface? The crucial questions are What is a whale? What is a fish? In the absence of knowing, we should not treat these animals as if they were inorganic particles moving around randomly.

At an even more basic level, we are coming to one of the really fundamental questions. Until now, we have been living with the idea that we can predict. If we could not predict, we assumed that with more data and better models we would be able to do so. It is now becoming quite clear at the level of ecological theory that the complexity of natural systems makes them essentially unpredictable. There are some complex systems whose behavior we will never be able to predict. But we can theoretically determine some of the answers we need and measure their uncertainty; we do not have to say simply that they are uncertain. I think we are

going to have to choose courses of conservation action considering those measures of uncertainty. Our whole approach to management will come to be based on probabilities.

QUESTION. A whale is affected by oil in one way; man finds oil unpleasant to look at. Both seem like fair factors to incorporate into environmental models. Were you saying that scientists and decision makers should not think about the way we react to oil? It seems to be an important factor.

ANSWER. A quite extraordinary thing happens in fisheries research. It explains much of the apparent trouble with predicting human behavior. Countries that have traditions of free enterprise fear discussions of economic control in international political forums. They more than fear them, they attempt to suppress such discussions. One consequence is that social analysis has been terribly impeded. I will give just one example. In the Northwest Atlantic Fisheries Commission, we made calculations for twenty years about what would happen if we changed the mesh size of trawler nets. We were forced by the politicians to assume not just that the fish were particles milling about, but that the fishermen were stupid and placid. We were not allowed to put assumptions into our models about the way fishermen would react to what we proposed to do. That ruined the predictive value of the analytical system, but the system was still used. This is one of the reasons there are virtually no haddock off Georges Bank, no anchovies off Peru, and no herring in the North Sea.

We have to build not just ecological models, but social models, with all their uncertainties. We know something about the way humans behave, and we have to try to take that information into account.

QUESTION. Questions of ocean and river policy seem to be far beyond the influence of individual citizens. Isn't it fair to say that this is one area where local action is just not possible? Perhaps you can try to influence the national leadership, but that is not a very direct solution.

ANSWER. There is much more individuals can do. As you know, the Japanese government has a very bad record on many environmental issues. The opposition to their policies comes directly from the people in most cases. Minamata is the classic case.

People were dying from mercury poisoning as a result of eating fish contaminated by the effluent of Japanese factories. When people realized what was happening, there was serious confrontation between government and industry on one side and citizens' organizations on the other. At times it was very violent. As a result, Japan enacted strong legislation against many kinds of environmental pollution. This led their big corporations simply to export polluting industries to other, less developed countries in Southeast Asia. The Japanese nongovernmental organizations that had forced the domestic change began a concerted action to stop this practice within a year. This was a superb demonstration of global conscience; unfortunately no other nation has duplicated it. The Japanese NGOs sent commissioners abroad to show people how to fight the Japanese multinational corporations that were bringing polluting industries to their countries. This is, I think, a fantastic example for NGOs in other countries.

You can start doing things to help control the actions of corporations from your own country in other parts of the world. This is the best first step toward effective international action.

Food, Nutrition, and Population

JEAN MAYER

Except for nuclear war, the greatest threat to the environment is the explosive growth of our human population. All other threats are minor compared with the problems of caring for an ever-larger number of people. The situation is best exemplified by an old French riddle: "If the lilies in a pond are doubling the number of their leaves every day and the pond is full at the end of the thirtieth day, when was it half full?" The answer is, "On the twenty-ninth day." And on the twenty-ninth day everyone still feels that there is plenty of room in the pond.

Population Size

Early historical data on population size are not completely reliable, but we do have reasonably accurate information from the seventeenth or eighteenth century to the present. The oldest data we have come from the beginning of our era. People forget that Jesus was born in Bethlehem because a large census was being conducted in the Roman Empire and everyone was going back to their ancestral homes to be counted. We have the data from that census. We also have some data that were collected at the same time in China. Those were the two large population centers of the period. The population of the Roman Empire was estimated to be 57 million, that of China 50 million. As far as we can estimate, the population of the world 2,000 years ago was clearly less than 200 million people. The population of the countries that had been in the Roman Empire went up rapidly during the 400 years of peace that the empire secured for western Europe and the Mediterranean. It went down again at the time of the great epidemics that accompanied the breakdown of the empire's defenses and

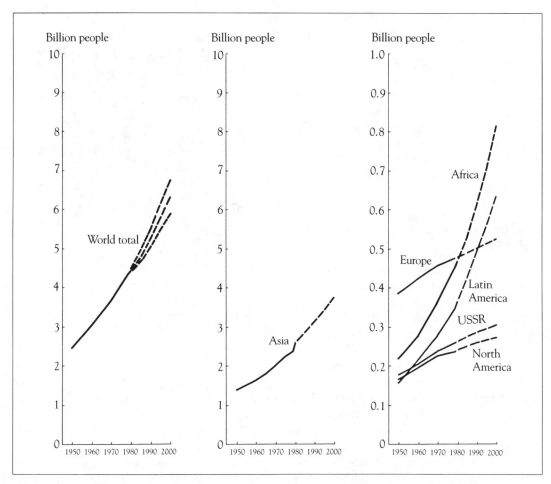

FIG. 14.1. *World population by region, 1950–1979, with projections to 2000. Most of the increase is expected in developing countries. Africa and Latin America are projected to double in population. Asia will continue to have the largest number of people. Graphs from CEQ.*

the barbarian invasions. It went up and then down during the Middle Ages as the great plague killed off one-third of the population of Europe in less than one year. The populations of China, India, and South America showed similar ups and downs. As far as we can tell, world population did not reach 500 million until the year 1500. From then on, however, progression was rapid. It took only 350 years to double to 1 billion in 1850, 70 years to double again to 2 billion by 1920. World population reached 3 billion in 1965, 4 billion in 1977. There are now about 4.5 billion of us in the world (United Nations, 1981b).

The rate of population increase is slowing. In the early 1960s the projected rate was 2.2% per year, which would have meant about 8 billion people by

the year 2000. The current estimates, based on a growth rate of 1.7%, suggest that there will be about 6.3 billion people by the year 2000 (United Nations, 1981b). Eighty percent will be living in the developing world, where over 90% of the increase will take place. At present, the best guess (and population rate guesses are notoriously inaccurate) is that fertility will continue to decline worldwide. We should reach a replacement level of reproduction by the year 2020. If this occurs, the population of the world will stabilize at about 11 billion. If we are extraordinarily successful and work extremely hard on population problems, and if we reach this replacement level by the year 2000—which is probably more than we should hope for, and certainly more than we should expect—the population will plateau at roughly 8 billion, about twice the present number. Obviously, avoiding a gain of 3 billion people by the end of the twenty-first century would be a tremendous boon from the points of view of environmental preservation and the quality of the lives of our great-great-grandchildren.

Food, Income, and Population

Most discussions of the problems of food and population are replete with oversimplifications. H. L. Mencken once said, "For every complex and difficult problem there is a simple and obvious solution—and it is wrong!" That statement applies over and over again in this area. One common oversimplification is to link the problem of feeding people to the problem of overcrowding. The Netherlands and Belgium are much more crowded than India, but people in these countries are not starving. The one simple statement I would dare to make is that no matter what the climatic conditions, no matter what the degree of crowding or the degree of development, wealthy people have enough to eat. India has a per capita gross national product (GNP) of U.S. $180, while the Netherlands has a per capita GNP of U.S. $8,390 (World Bank, 1982). This explains why the Dutch are well fed and the Indians are struggling.

It is true that developing countries have a much faster rate of population increase than developed countries, but we cannot generalize from that to say that wealthy people have fewer children. At present, Kuwait has the highest rate of population increase, 3.9% per year, but it also has one of the highest per capita incomes in the world, U.S. $12,000 per person per year (World Bank, 1982). If you look at other developing countries, such as Somalia, Venezuela, or Brazil, you find no clear correlation between wealth and the rate of population increase. The notion that all we need to do is to increase income and the birthrate will fall auto-

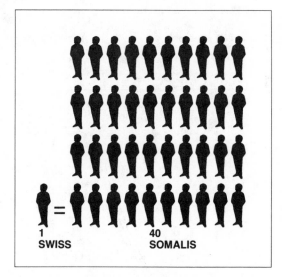

FIG. 14.2. *Disproportionate consumption of resources by the affluent: One Swiss consumes as much as forty Somalis. Illustration from IUCN, 1980.*

matically is clearly false. Cultural factors, and the time it takes to change cultural factors, are much more important determinants than economic factors.

Overpopulation is a relative term. It tends to connote misery rather than density. People always talk about India as overpopulated, but never mention Denmark, which in fact has a much higher population density. The presence of more people is serious, however, even in developed countries, because of the stress they put on the environment and the air and water pollution they cause. The relationship between numbers of people and wealth is also important in terms of competition for goods and services. Developed market economies import about 75% of the world's primary products, including food, fuels, and crude materials. They also import 65% of the world's manufactured goods (Hansen, 1982). More people, especially wealthy people, deplete resources more quickly.

Crowding, on the other hand, is important in terms of the spread of environmental illnesses, particularly the mass diseases of mankind—schistosomiasis, malaria, typhoid, diphtheria, poliomyelitis. Few people understand that a sparse population is better protected from the spread of infectious dis-

FIG. 14.3. *Mount Kilimanjaro, which should be clearly visible in this photograph, is almost totally obscured by smoke from brush fires. Photograph by E. Garrety from Massachusetts Audubon Society.*

eases. Careful studies have been made of the life expectancies of people living in France about 10,000 years ago based on skeletons that we have uncovered. It is obvious that life expectancy was very good at that time. There do not appear to have been epidemics among children. It is likely that the population was so sparse that diseases with epidemic potential could not spread very far. The people who lived along the lakes of France 10,000 years ago probably had a comfortable and healthy life compared with the inhabitants of crowded early-nineteenth-century cities, or even with many of the urban poor today.

The Human Environment

The enormous increase in global population projected for the end of this century may also cause climatic changes. There will be more carbon dioxide, airborne dust, and thermal pollution from the cities. In developing countries smoke from the constant burning of brush to clear land for agriculture will probably have the most destructive environ-

mental impact. Scientists are now considering whether the carbon dioxide produced by slash and burn agriculture and the burning of fossil fuels will trigger massive changes in the environment, and if so whether it will cause a cooling or warming trend. It will probably be 1995 before we have data that will give us the answers.

Finally, there is little doubt that population growth causes a huge and unsatisfied demand for all kinds of basic services. For example, in Kenya there is one doctor for 50,000 people. In Thailand, which is not a terribly poor country, public health services reach only 17% of the population (United Nations, 1981b). The problem of housing the gigantic conurbations (aggregations of urban communities) that are spreading in developing countries should cause anxiety. Some cities are becoming enormous, far larger than any that have ever existed before. Mexico City may have over 20 million residents before the end of the century: Lagos may have 8 to 10 million; Calcutta may have 15 to 20 million. There is no evidence that the funds exist to house these people. Yet they keep piling into shantytowns and never go back to the countryside. Notions of life in the country may be romantic, but people just do not want to live in villages. They want to escape what Lenin called the stultifying idiocy of village life. As bad as it is to live in a shantytown, if your child is sick there you are much more likely to get her to the hospital. You are more likely to have your child learn to read and write, to use better water, to see a movie once in a while. You cannot have any of these things in most of the villages of the world.

The increase in world population also puts stress on educational facilities, even though almost all developing countries are conscious of the need for literacy. In 1950 there were 700 million illiterate people in the world—44% of the adult population. Today the percentage has shrunk to 34%, but in fact this represents 800 million illiterates (Hansen, 1982). There is almost no statistic that better illustrates how population growth can nullify the enormous development effort that has been made.

Further, population increases have a profound impact on employment. In the United States we have an unemployment rate that fluctuates around 9%. We recognize that 1% to 3% may be an irreducible minimum. The International Labor Organization estimates that about 29% of the population in developing countries is unemployed (World Bank, 1982). They also estimate that for every 1% growth in the labor force we need 3% economic growth. That is a very difficult multiple to achieve. A country that has 3% population growth needs a 9% rate of economic growth to maintain roughly the same rate of employment.

Pressure on the physical environment and the food-sustaining systems arises from three general categories of problems. First, and obviously, population growth increases the number of consumers. If you have twice as many people, they will eat twice as much.

Second, as individuals within a population become more prosperous, their consumption of ani-

FIG. 14.4. *This Nepalese man is being carried to a health clinic by his son. Others, who live even farther from the clinic, have no access to it at all. Photograph by M.J. Odell.*

mal products increases. This entails an increase in the per capita use of grain. The most modern methods of producing milk and eggs (the animal products most efficiently converted to calories) only have a conversion rate of 25%. That means you get back 25% of the calories you feed to very efficient milk cows or laying hens. For other animal foods, the return is even lower. The production efficiency of broiling chickens is about 18%, of turkeys about 16%, of hogs about 12%, and of beef from feedlots about 6% or 7%. Ninety percent of the grain we consume in the United States is first fed to animals. Only 10% feeds people directly (Brown, 1978). By comparison, the largest amount of grain that Bangladesh ever imported in one year was about 3 million tons in 1973, roughly the same amount we use each year for beer manufacturing. Prosperity means that enormous amounts of food are put to indirect use. It increases the pressure on both the food supply and the food-supporting ecosystems.

The third factor is the increasing stress the human population places on ecosystems themselves. Croplands throughout the world are beginning to suffer from erosion, overfertilization, overirrigation and resulting salinization, urban encroachment, and the strangest threat to the environment anyone has yet thought of—the manufacture of gasohol, the transformation of grain to alcohol in a very inefficient fashion. To create gasohol, petrochemicals are used for power, fertilizers, and pesticides. Then more oil is used to distill the alcohol after fermentation and to recover it. This makes no sense from the point of view of energy conservation, because about the same amount of energy is consumed that is produced. The only virtue of gasohol is that it may give a little energy independence because you can burn coal and lignite as substitutes for oil during some of the steps in its manufacture.

The manufacture of gasohol also has several other disturbing implications. If you do not increase the land under cultivation, then you are setting up an investment system that will create havoc during a bad year for agriculture. When we need all the grain we have to feed the world, we will be running our cars with it instead. The only possible solution would be to cultivate marginal lands. This requires much more power, more pesticides, and more fertilizer. It also increases erosion. It makes some sense to convert biomass other than grain into alcohol, as the Brazilians are doing. For example, we may want to convert wood chips and sawdust into ethyl alcohol. But it makes no sense to convert grain. In the 1970s a hectare of arable land supported an average of two people. By the year 2000 a hectare will have to support four people—and it is not going to do so by producing gasohol.

The erosion of arable land is a profound prob-

lem. The United Nations estimates yearly global losses at 3.5 million hectares of rain-fed cropland and 125,000 hectares of irrigated cropland (Brown, 1978). In developed countries the problem is made worse by urban encroachment. Cropland is very much underpriced in the United States. It is too economical for developers to transform croplands into supermarkets, airports, suburban housing, and the like. Because of development one-sixth of the cropland in California has disappeared since 1945. It is projected that there will be no cropland left in Florida by the year 2000 (Brown, 1978).

Dry grassland ecosystems are suffering from overgrazing and desertification, not because of changes from past practices but because there are more animals and more people. Their current rate of utilization allows no time for renewal. The southern border of the Sahara is advancing at a rate of about 1.5 miles per year, and the same is true in many other desert areas around the world.

Croplands and grasslands are two of the crucial ecosystems under threat; the third is fisheries. The world fish catch peaked in the 1970s at about 70 million tons and is now slowly moving down in spite of, or perhaps because of, the continued increase in the size of fishing fleets (Brown, 1978). Pollution from oil spills is an additional problem. If we stay at the present level, and there is no guarantee that we can maintain it if overfishing and spills continue, those 70 million tons of fish should be enough to supply 27% of the protein requirement of 4 billion people—if all the fish are used as food. But most of the fish used in this country go for the production of poultry feed and fertilizer. Unfortunately, we expect the catch to fall to about half the present level by the year 2000. Fish will represent an ever-decreasing proportion of the world's protein requirement.

Finally, there is the problem of wood supply. The only area in the world replacing its forests is the northwest United States and Canada. In the southeast United States forestry is increasing, and the Northeast has a problem with acid rain, as does Sweden. About 90% of the wood cut in developing countries is used for fuel. The greater demand for cooking fuel from an increased population means that Latin America, most of Asia, Africa, and

Oceania will lose about 40% of their present forests by the year 2000 (Brown, 1978).

The Outlook
for the Future

As I suggested earlier, the problem of feeding the world is extraordinarily complex. It not only has to do with the production and distribution systems, but also with buying power. There have been some encouraging success stories. India, for example, has made enormous strides in the last decade. They have not imported food for three years. In fact, they have been able to export to Cambodia and Indonesia and they have roughly 25 million tons of grain reserves (United Nations, 1981a). This seems incredible compared with India's situation in 1973. But there are still many people in India who are poorly fed, and the country still has a shaky economic system, difficult distribution, and too few jobs. There can be surpluses in one area and deficits 30 miles away. But despite these distribution problems, the Indians have done very well.

Their success was based on the techniques of the green revolution—the use of new crop varieties that respond to petrochemicals. It required an increase in the use of these petrochemicals for soil preparation, irrigation, fertilizers, pesticides, the building of storage facilities, and transportation. However promising we find other techniques, and we should make concentrated efforts to develop them, it is going to take at least twenty years for the next generation of agricultural production methods to make an impact on the world. In the near future we must face the dilemma that if we are going to feed the world, we are also going to accelerate the use of petrochemicals. There is no alternative.

Obviously, technological advances are not enough to increase food production. There must be effective demand for the food that is produced. A farmer is not going to grow more unless he knows he can sell it. This means that the people in those vast conurbations must have jobs. Creating jobs in the cities is, in many ways, the best method for increasing agricultural production.

Food aid, by contrast, is a very difficult tool to use. Most of Africa will become dependent on the United States for food in the years to come. Turn-ing this around will entail changes in the social and economic structures of the recipient countries. It can be done, but not immediately. And it has to be done carefully so that the threat to their environments is not increased.

We must begin to think of the problem of food production in three different scales of time. We urgently need an early warning system for famines, probably one based on indices of weather and economic and medical data. (Sadly, one of the best indices of famine is the growth rate of children in populations at particular risk.) We also need a system that can respond to the warnings very quickly, which means we need a ready reserve of food. Most logically, this would be in ships at sea, which at any one time are carrying millions of tons of grain to countries with sufficient buying power. These ships could be diverted to the endangered area, as was done during the famine in Bangladesh. We also need reserves that can back up those shipments in both producing and recipient countries. Of course, all kinds of food aid other than emergency aid for famines have to be used with some discretion, because the last way to increase production by the farmers of the world is to dump free food on their market.

We also need massive technological assistance in agriculture, so that in the longer run those needy lands where the greatest population growth is taking place can become self-sufficient in food. The United States has a special role to play here. We are the people who invented the land-grant colleges, agricultural extension, and rural credit, the essential support systems of modern agriculture. We are very good at food production and should share our expertise. Finally, we need intensive research to expand the kinds of traditional foods that are available and to develop entirely new foods.

All this needs to be done with an eye to conservation, so that we do not destroy the ecosystems that are the foundation for production not only for the human species but also for the multiplicity of species for which we are stewards. It will not be easy. I see no signs of a development of the consciousness we will need in the power centers of the world. In fact, we are in the midst of an enormous regression of assistance to poor countries. There was more concern—and action—twenty or thirty years ago. It is probably the twenty-ninth day at four o'clock in the afternoon, and very few seem to be aware that the pond will soon be full.

Discussion

QUESTION. How confident are you that population growth will decline to a replacement rate by the year 2020? What is causing this decline?

ANSWER. I am not confident about it at all. This is a projection of the U.N. Population Bureau. In my lifetime I have seen demographers' projections proven wrong over and over again, although they did project the population boom that occurred after World War II. I remember the time in the late 1930s when the disappearance of the white races was the great cry of alarm among the people of Europe and the United States. Many were worried that the developed countries would be swamped by explosive population growth in the Third World. That boom came, but we are still here. Very large changes in population characteristics have seldom been predicted properly by the demographers. Still, the figure I quoted is the best we have.

At present, there is a decline in the overall rate of population increase in the world. One hundred governments have proclaimed that population stabilization is one of their aims. This situation is very different from the past. In the mid-1970s, the African Bureau of the World Health Organization was refusing to abide by central directives from Geneva to make birth control available to local people. Many Third World people thought that birth control in Africa was a racist plot; this was a holdover from the 1930s mentality I mentioned. Opposition to birth control was even then the official position of several African governments. That sentiment is essentially gone now. Most of the Latin American countries have agreed that birth control information is important. China has stated that it will reduce its population growth rate by 50%. So it is not so terribly absurd to think that an overall decline in world population growth will continue.

QUESTION. Why is it that population increase has caused famines in some impoverished areas and not others? There does not seem to be a simple relationship between the number of people and the effective production of food as causes of famine, although it seems that there should be one.

ANSWER. Let me dispel one of the great misconceptions about the relationship between food and population. There is a mistaken belief in some quarters that an increase in population leads to famine. So, this line of thought continues, as the world becomes more populated, we should have more famines. That is not likely. I have had the melancholy experience of coping with several famines in my professional life. I was involved with famine relief efforts in the Congo, Bangladesh, and in Biafra. I can assure you that when there is too little food, the one thing that does not take place is a further population increase. To the contrary, women abort, babies do not survive, the elderly become clearly less resistant to disease. People die. Famines are a drastic check on population.

The real result of population increases is a progressively more mediocre life, an increasing shortage of decent food, more malnutrition, more environmental damage, greater susceptibility to a variety of diseases. The problems of food and population have to be séen in that light, not as a cause of sudden dramatic famine.

QUESTION. Is there any real evidence that more energy goes into producing food with modern methods than actually comes out, and, if so, what does this suggest?

ANSWER. One of my associates and I have published calculations on the cost of producing food in the United States. On the average, it takes 10 calories of petrochemicals to produce 1 calorie of food for Americans. Agriculture used to be a way of transforming sunlight into food, now it is a way of transforming oil into food. As for the implications, I think it is sufficient to say that no system can operate with a net energy deficit over the long run.

QUESTION. As a biologist specializing in studies of behavioral ecology, I have been puzzled for a long time by the inability of the human species to control its population growth. It has occurred to me recently that cities might act as a major factor in people's inability to gauge their own population rate as animals do. When a habitat is fully occupied by animals, the offspring have to migrate to find their own place. Natural mortality is much higher because if there is no place to support the animals, they die. With the human species the overflow dispersal from the country is moving to the cities, so there is no natural control of the growth rate. Will

the cities continue to be the "storage" for our surplus population?

ANSWER. You can push that line of thought further. It is possible to think that the earth is an organism and that the human species is a cancer with
unregulated growth. In this model, multiplication
is affected by controlling factors both inside and
outside the population. You are suggesting that
some of the signals have been affected by the presence of cities.

There is, however, a fundamental problem with
this model. It entirely ignores the dignity of individual people. It is critical that we control our
numbers, but it is equally essential that individuals
are never thought of as "surplus" and therefore
without value.

V Conclusion

Toward Sustainable Development

JEAN-CLAUDE FABY

When environmental problems were first discussed internationally in the early 1970s, they were prone to be quickly identified, in popular perception, with the problems of pollution control, conservation of wildlife, and aesthetically motivated protection of nature. Certain misgivings were expressed by developing countries: They considered that concern about their environment was unnecessary and would jeopardize their efforts to industrialize rapidly and modernize their economies. More recently, misgivings have been expressed in certain developed countries as well. They believe that increasing national expenditures for environmental protection and improvement are inducing inflation, unemployment, and economic stagnation.

It has been clear to the U.N. Environment Programme (UNEP) from the outset that environmentally sound development policies and measures should have as major aims

1. Minimization of waste in the use of natural resources
2. Maximization of productive use of residues of all kinds
3. Respect for the integrity of ecosystems (modifications should be introduced only after careful evaluation of their likely consequences)
4. Minimization of the degradation of the environment, and maximum abatement of such degradation when it has been unavoidable
5. Harnessing, to the maximum extent possible, of the complementarities between environmental improvement and socioeconomic development.

Now that it is generally seen, and sometimes even accepted, that environmental and developmental objectives can be pursued in mutual support and, in fact, that they are interdependent, the next issue we must face is how such a complex relationship can be given an operational context. But it seems that to succeed in harmonizing these two sets of objectives, four basic prerequisites have to be fulfilled.

1. There must be appropriate methodological frameworks within which alternatives can be evaluated to determine both developmental and environmental benefits and their associated costs. Such methodological frameworks cannot be uniform or rigidly applicable in all countries. But there is no doubt that uncritical reliance on a narrowly conceived economic calculus of financial rates of return is inappropriate if the goal is to reach environmentally prudent and socially satisfactory solutions.

a. Cost-benefit analysis with regard to environmental management measures is gaining interest, although there is still concern about its precision. Some even take a cynical view of such studies. However, with the economic conditions faced by most countries, allocation of limited and scarce financial resources for competing societal needs becomes a major task.

The UNEP has undertaken the systematic review of over one-hundred case studies provided by member countries. Conclusions reached so far are that there is no one format or agreed definition for cost-benefit analysis of environmental management. Indeed, the approaches taken differ in the degree of concern expressed for environmental considerations, in their underlying scales for sociocultural values, in the depth and rigor of their research, in the scope of the system dealt with, and in the period of time that should be considered. They also differ from one country to the next—even within a country—and in the type of sectors studied. In sum, they vary widely in both form and content.

In spite of these substantial differences, the case studies do show a similar analytical approach. The great majority of them were based, with different degrees of success, on traditional economic methodologies, oscillating between microeconomic and welfare theory. This similarity would seem to stem from the lack of new techniques of analysis for environmental economics. Certain case studies reached the conclusion that, given the state of analytical tools and technical knowledge, a cost-effectiveness analysis is perhaps a more viable alternative than a cost-benefit analysis. At present, perhaps the best we can do is to look more narrowly at what the return will be for expenditures on environmental improvements. It is, however, possible to envisage situations where both tools could be used to complement each other. The experts have recommended development of a step-by-step methodological framework for carrying out cost-benefit analysis studies.

b. Risk management is emerging as an important component of environmental management. This topic relates closely to environmental impact assessment as well as cost-benefit analysis. There are a number of key issues involved, including the notion of voluntary and involuntary risks; identification of risks; tools and techniques for assessing and quantifying risks; and, most important and topical, risk acceptance. Who should decide that a particular level of risk is acceptable? Subjective value judgments, and indeed morality and ethics, are apparent in this basic issue.

c. Environmental impact assessment (EIA) is emerging as a major tool for incorporating environmental considerations in development projects. Environmental impact assessment, if used appropriately, and at the early project planning stage, is coming to be perceived as a useful part of the overall planning process. More important, the EIA procedure can generate alternatives for siting, processes, raw materials, products, and so on. Decision making is enhanced when the range of alternatives increases. Environmental impact assessment has received increased recognition since the Joint Declaration of Environmental Policies and Procedures relating to Economic Development was recently signed by UNEP, the U.N. Development Programme, the World Bank, and six regional banks. The Brandt Commission Report also stated, "Environmental impact assessment should be undertaken wherever investments or other development activities may have adverse environmental conse-

FIG. 15.1. *The environmental consequences of development are often considered after the fact. Construction on this road through the Monte Verde Cloud Forest Reserve in Costa Rica was eventually halted, but only after it had caused substantial damage. Photograph by G. Bertrand.*

quences whether within the national territory concerned, for the environment of neighbouring countries or for the global commons" (ICIDI, 1980).

d. Finally, increasing attention is being paid to the formulation of technical guidelines for incorporating environmental consideration in project planning and execution. The UNEP, the World Bank, the International Development Bank, and the U.S. Agency for International Development all have prepared or are preparing such guidelines for a variety of development sectors. We are currently attempting to ensure the wide dissemination of these guidelines to multilateral and bilateral development financing institutions.

2. A second prerequisite is appropriate institutional frameworks to facilitate decision making and action. Such institutional machinery encompasses environmental legislation as well as systems of incentives and disincentives.

a. At the national level, developing countries have followed and implemented or are beginning to implement procedures similar to those in the industrialized countries. For example, with regard to organization, most developing countries have established and staffed an office, department, or ministry for the environment. Some are elaborate structures with large personnel, while others are smaller. Laws related to the environment have generally been promulgated. Some involve the adaptation, amalgamation, or updating of existing laws and ordinances. Some are new laws, specifically addressing environmental protection. There is also a tendency to set ambient air quality, stream quality, and discharge standards. Incentives, in the forms of taxes and subsidies, are being used, and penalty and fine mechanisms exist for those contravening the pollution laws. Generally, however, there is a lack of enforcement and, of course, the relative clout of the environmental department or ministry varies widely from country to country, as does the level of sophistication of applicable standards.

b. At the international level a relatively important amount of work has been accomplished in the past ten years. Institutionally, the establishment of UNEP in 1972 has been followed, although relatively slowly, by the emergence of environmental mechanisms and procedures in a number of multilateral and bilateral institutions in development financing. Legislatively, some significant achievements took place in the 1970s. The Third U.N. Development Strategy includes, for the first time, references to environmental training in both its objective and policy measures sections. Paragraph 41 of the goals and objectives section reads:

Accelerated development in the developing countries could enhance their capacity to improve their environment. The environmental implications of poverty and underdevelopment and the interrelationships between development, environment, population and resources must be taken into account in the process of development. It is essential to avoid environmental degradation and give future generations the benefit of a sound environment. There is need to ensure an economic development process which is environmentally sustainable over the long run and which protects the ecological balance. Determined efforts must be made to prevent deforestation, erosion, soil degradation and desertification. International cooperation in environmental protection should be increased.

Although couched in very general terms, this includes most key elements of a sustainable development policy. The policy measures subsection dealing with environment reads:

156. Because health, nutrition and general well-being depend upon the integrity and productivity of the environment and resources, measures should continue to be developed and carried out to promote the environmental and ecological soundness of developmental activities. Methods will be devised to assist interested developing countries in environmental management and in the evaluation of the costs and benefits, quantitative and qualitative, of environment protection measures with a view to dealing more adequately with the environmental aspects of development

activities. This method will be developed taking fully into account the existing knowledge of interrelationships between development, environment, population and resources. To that end, research on these interrelationships will be intensified. The capacity of the developing countries will be strengthened to facilitate their making appropriate scientific and technical choices relating to environment in their development process.

157. Bilateral and multilateral donors will consider, within the over-all financing of projects in developing countries, at their request, meeting the costs of taking environmental aspects into account in the design and completion of such projects. They will furthermore provide assistance, including in the field of training, to develop the endogenous capacity of developing countries to follow the methods enumerated in paragraph 156 above, thereby also facilitating technical cooperation among developing countries.

158. The International Community, in particular the developed countries, will substantially increase its financial and technical support to drought-striken countries suffering from desertification. In this context, support for the Plan of Action to Combat Desertification will be augmented.

A number of other points made in the Third Development Strategy have a built-in environmental perspective; for instance, the energy sections have a strong conservation and renewable resources orientation. There are also sections on health, human settlements, and natural disasters in which environmental aspects are evident.

Another significant development has been the coming into force of a number of new international instruments dealing with environmental concerns, for example, the Convention on Trade in Endangered Species, the Convention on the Protection of Wetlands of International Importance, and the World Cultural Heritage Convention. In Europe an important development was the adoption of the Convention on Long Range Transboundary Pollution under the auspices of the U.N. Economic Commission for Europe in 1979. In addition, the Regional Seas Program of UNEP has led to the adoption of a series of international instruments for, so far, the Mediterranean, the Persian-Arab Gulf, and, very recently, the Gulf of Guinea. Finally, work in the past few years toward a new law of the sea treaty contains specific environmental provisions. All these developments point to an

FIG. 15.2. *Large developments such as this Egyptian water project run the danger of ignoring the attitudes and knowledge of the people they are intended to help. Photograph by M.J. Odell.*

emerging body of international conventional law dealing with environmental concerns.

The U.N. Environment Program's work in this area, which was somewhat fragmented, was firmed up and expanded following a major meeting of experts in 1982, which reviewed a long-term program of action for UNEP in the area of environmental law, based on an analysis of regional priorities.

3. Another important requirement for sustainable development is the formulation of concrete alternatives for technologies, processes, practices, and products in terms of both production and consumption. These should result in economies in the use of scarce natural resources and minimize environmental degradation. They also should maximize the chances of simultaneous attainment of development objectives—such as increased production, expanded employment, and eradication of poverty— and environmental objectives—such as improvement of environmental sanitation and health and prevention and control of deforestation, desertifi-

cation, and soil erosion. Obviously, implementation of such alternatives may in some situations entail definite and far-reaching changes in life-styles and consumption patterns.

One major key to sustainable and socially satisfactory development is the adoption of environmentally sound and appropriate technologies in all fields. This concept has sometimes led to uncalled-for apprehensions and unintended interpretations, so it is important that it be correctly understood.

People have often assumed that the proponents of environmentally sound and appropriate technologies demand a total rejection of the "modern" technology of the developed countries. In fact, what is demanded is a careful scrutiny of the economic, social, and environmental implications of such technology. In some quarters, the argument for environmentally sound and appropriate technologies has been misunderstood as a plea for a return to, and total dependence on, the traditional technologies of ancient peoples. In fact, the plea is quite different. Traditional technologies have undergone a selection process over centuries of empirical testing, hence they are very likely to represent optimum solutions. But they are optimum only for the particular conditions, constraints, materials, and needs for which they were developed. With the emergence of new conditions, their applicability is usually eroded and they are rendered invalid. Nevertheless, it is quite possible for these traditional technologies to undergo qualitative changes through minor modifications that use modern science and engineering to clarify the rational core of ancient practices.

What is environmentally sound and appropriate at one juncture may not be so at another. The concepts of environmental soundness and appropriateness are not static; they must evolve with the state of the environment and with the nature of development tasks.

4. A fourth prerequisite for making environmentally sound development a reality lies in the attitudes, perceptions, information, and knowledge of the people themselves about what is environmentally prudent and in their best interests. Dissemination of environment-related information and environmental education and training must play a valuable role in bringing about an alliance between popular and technical evaluations of various alternatives for development.

This requirement is hardly one that I need to emphasize here. Nonetheless, it is one of the most fundamental needs of countries, developing countries in particular, if their capacity to perceive environmental stress and manage their natural resources bases is to be enhanced.

There is no doubt that the entire world stands to benefit from a more rational use of physical and human resources. The developing countries need more development assistance to learn from past environmental mistakes of industrialized countries and to realize the environmental opportunities they have. But availability of environmentally prudent and developmentally satisfactory technology is not enough to improve the state of the earth. To keep the dangers at bay and make use of promising opportunities, economic and institutional reforms in international relations have to occur. If only a real relaxation of tension between the East and the West and genuine belief in self-reliance and in interdependence between developed and developing countries can be achieved, the prospects for improving the quality of life everywhere will be greatly enhanced.

Abbreviations

CEQ	(President's) Council on Environmental Quality
FAO	U.N. Food and Agriculture Organization
ICEE	Intergovernmental Conference on Environmental Education
ICIDI	Independent Commission on International Development Issues
IFDA	International Foundation for Development Alternatives
IUCN	International Union for Conservation of Nature and Natural Resources
UNEP	U.N. Environment Programme
UNESCO	U.N. Educational, Scientific, and Cultural Organization

Bibliography

Aldrich, James L. (1977) *Trends in Environmental Education*. Paris: UNESCO.

Allen, G. M. (1942) *Extinct and Vanishing Animals of the Western Hemisphere*. Special Publication No. 11. Washington, D.C.: American Committee for International Wild Life Protection.

Allen, W. (1965) *The African Husbandmen*. Westport, Conn.: Greenwood Press.

American Society of International Law. (1977) *A Global Satellite Observation System for Earth Resources: Problems and Prospects*. Washington, D.C.: West Publishing.

Artha Shastra (c. 300 B.C.) Quoted by the secretary general, Indian Board for Wildlife, Pan Indian Ocean Science Foundation, 1952.

Asibey, E. O. (1969) "Grass Cutter *Thryonomys swinderianus* as a Source of Bushmeat in Ghana." Accra, Ghana: Department of Game and Wildlife.

———. (1974) "Wildlife as a Source of Protein in Africa South of the Sahara." *Biological Conservation* 6:32–39.

Asoka. (c. 250 B.C.) Pillar Edicts. Quoted by Maharajah of Mysore in *Indian Wildlife Bulletin*, Dec. 1952.

Bader, Richard, R. A. Rogotzkie, and J. M. Teal. (1972) "Transportation and Coastline Modifications." In *The Water's Edge*, ed. by Bostwick H. Ketchum, pp. 125–145. Cambridge: MIT Press.

Barr, T. N. (1981) "The World Food Situation and Global Grain Prospects." *Science* 241:1087–95.

Bennett, Hugh H. (1939) Quoted in testimony of Peter R. Huessy, before the Subcommittee on Environment, Soil Conservation, and Forestry of the U.S. Senate Agricultural Committee, Aug. 2, 1977.

Benson, W. W. (In press) "Amazonian Ant-Plants." In *Amazonian Rain Forest*, ed. by G. T. Prance and T. E. Lovejoy. Oxford: Pergamon Press.

Brightbill, C. K. (1960) *The Challenge of Leisure*. Englewood Cliffs, N.J.: Prentice-Hall.

Brown, L. R. (1978) *The Twenty-ninth Day*. N.Y.: W. W. Norton.

———. (1981) "World Population Growth, Soil Erosion, and Food Security." *Scientific American* 213:995–1002.

Brown, S., et al. (1977) *Regimes for the Ocean, Outer*

Space and Weather. Washington, D.C.: Brookings Institution.

Burger, George V. (1978) "Agriculture and Wildlife." In *Wildlife and America,* ed. by H. P. Brokaw, pp. 89–107. Washington, D.C.: CEQ.

Buringh, P. (1977) "Computation of the Absolute Maximum Food Production of the World." (Cited in Hanson, 1979).

Central Intelligence Agency. (1978) *Polar Regions Atlas.* Washington, D.C.: GPO

Coale, A. J. (1974) "The History of the Human Population." *Scientific American* 231(3):41–51.

Coe, M. (1980) "African Wildlife Resources." In Soule and Wilcox (1980), pp. 273–302.

Commonwealth of Australia, Department of Home Affairs and Environment. (1982) *National Conservation Strategy for Australia: Living Resource Conservation for Sustainable Development.* (Towards a National Conservation Strategy—A Discussion Paper).

Connect [UNESCO-UNEP Environmental Education Newsletter]. (1976) Vol. 1, No. 1.

Convention on Wetlands of International Importance, Especially as Waterfowl Habitats. (1972) Ramsar, Iran: n.p.

Council on Environmental Quality. (1980) *Biological Diversity.* Washington, D.C.: GPO.

———. (1981) *Environmental Trends.* Washington, D.C.: GPO.

Council on Environmental Quality and Department of State. (1980) *The Global 2000 Report to the President,* 2 vols. Washington, D.C.: GPO.

———. (1981) *Global Future: Time to Act.* Washington, D.C.: GPO.

Courtenay, W. R., Jr. (1978) "The Introduction of Exotic Organisms." In *Wildlife and America,* ed. by H. P. Brokaw, Washington, D.C.: CEQ.

Dadzie, K.K.S. (1980) "Economic Development." *Scientific American* 243(3):59–65.

Dasmann, Raymond F. (1964) *African Game Ranching.* London: Pergamon Press.

———. (1968) *Conservation and Rational Use of the Biosphere.* U.N. Economic and Social Council, E/4458, N.Y.

———. (1973a) "Classification and Use of Protected Natural and Cultural Areas." IUCN Occasional Paper 4. Morges: IUCN.

———. (1973b) "A System for Defining and Classifying Natural Regions for Purposes of Conservation." IUCN Occasional Paper 7. Morges: IUCN.

Dasmann, Raymond F., John Milton, and P. Freeman. (1973) *Ecological Principles for Economic Development.* London: John Wiley.

Dunne, T., and L. B. Leopold. (1978) *Water in Environmental Planning.* San Francisco: Freeman.

Eckholm, Erik. (1975) *The Other Energy Crisis: Firewood.* Worldwatch Paper 1. Washington, D.C.: Worldwatch Institute.

———. (1976) *Losing Ground: Environmental Stress and World Food Prospects.* New York: W. W. Norton.

———. (1979) *Planting for the Future: Forestry and Human Needs.* Worldwatch Paper 26. Washington, D.C.: Worldwatch Institute.

Ehrlich, P. R., and A. H. Ehrlich. (1970) *Population, Resources, Environment.* San Francisco: Freeman.

———. (1981) *Extinction.* New York: Random House.

Ejiogu, C. N. (1972) "The Kenya Programme: Policy and Results." In *Population Growth and Economic Development in Africa,* ed. by S. H. Ominde and C. N. Ejiogu, London: Heinemann.

European Common Market (1975) "On Pollution Caused by Certain Dangerous Substances Discharged into the Aquatic Environment of the European Community." Directives of the European Common Market.

Farnsworth, N. R., and R. W. Morris. (1976) "Higher Plants: The Sleeping Giant of Drug Development." *American Journal of Pharmacology* 146:46–52.

Farvar, M. Taghi, and John Milton. (1972) *The Careless Technology.* New York: Doubleday, Natural History Press.

Fearnside, P. M. (1979) "Cattle Yield Predictions for the Transamazon Highway of Brazil." *Interciencia* 4(4):220–225.

Federox, E., and I. Novik. (1977) "Man, Science, and Technology." In *Ecosocial Systems and Ecopolitics,* ed. by K. W. Deutsch, Paris: UNESCO.

Felger, R., and M. B. Moser (1973) "Eelgrass (*Zostera marina* L.) in the Gulf of California: Discovery of Its Nutritional Value by the Seri Indians." *Science* 181:355–356.

Fensham, Peter J. (1977) *A Report on the Intergovernmental Meeting Held in Tbilisi,* USSR. Paris: UNESCO-UNEP International Environmental Education Program.

Fontaine, J. P. (n.d.) "La Grande Mathematicomanie." *En Grande* 6(18):10–15.

Gadgil, M., and S. N. Prasao. (1978) "Vanishing Bamboo Stocks." *Commerce,* 17 June 1978.

Gorove, Steve (1973) "The Geostationary Orbit: Issues of Law and Policy." *American Journal of International Law* 1(1):5–12.

Goulding, M. (1980). *The Fishes and the Forest,* pp. i–

xii and 1–280. Berkeley: University of California Press.

Government of Canada. (1980a) *Arctic Pilot Project (Northern Component)*. Federal Environmental Assessment Review Office, Report of the Environmental Assessment Panel, No. 14, October. Ottawa.

———. (1980b) *The Lancaster Sound Region: 1980–2000 Perspectives and Issues on Resource Use: Executive Summary*. Ottawa.

Greenland, D. (1975) "Bringing the Green Revolution to Shifting Agriculture." *Science* 190:841.

Greenway, J. C., Jr. (1967) *Extinct and Vanishing Birds of the World*. New York: Dover Publications.

Haffer, J. (1969) "Speciation in Amazon Forest Birds." *Science* 165:131–137.

Hansen, R. D. (1982) *U.S. Foreign Policy and the Third World: Agenda 1982*. New York: Praeger.

Hanson, H. (1979) "Biological Resources." Report prepared for the Agricultural Production: Research and Development Strategies for the 1980's Conference. Unpublished.

Hardin, Garret. (1968) "The Tragedy of the Commons." *Science* 162:1243–48.

Harper, F. (1945) *Extinct and Vanishing Mammals of the Old World*. Special Publication No. 12. Washington, D.C.: American Committee on International Wild Life Protection.

Harroy, Jean Paul. (1971) *United Nations List of National Parks and Equivalent Reserves*. IUCN Publications, New Series, No. 15. Morges: IUCN.

Hatheway, W. H. (1971) "Contingency Table Analysis of Rain Forest Vegetation." In *Many Species Populations, Ecosystems, and Systems Analyses*, ed. by G. P. Patil, E. C. Pielou, and W. E. Waters, (Statistical Ecology, Vol. 3). University Park: Pennsylvania State University Press.

Hauser, P. M., and R. W. Gardner. (1980) "Urban Future: Trends and Prospects." (Cited in Brown, 1981).

Hecht, S. B. (1979) "Spontaneous Legumes of Terra Firma Pastures and Their Forage Potential." In *Increasing Forage Production on Acid Soils of the Tropics*, ed. by Sanchez and L. Tergas, Cali, Colombia.

Hollick, Ann L. (1979) "Oceans Regime for the 1980's." In Uzi B. Arad (ed.) *Sharing Global Resources*, pp. 171–203. New York: McGraw Hill.

———. (1981) *U.S. Foreign Policy and the Law of the Sea*. Princeton: Princeton University Press.

Honegger, R. E. (1981) "List of Amphibians and Reptiles Either Known or Thought to Have Become Extinct since 1600." *Biological Conservation* 19:141–158.

Hruby, T. (1981) "Shellfish Resources in a Polluted Tidal Inlet." *Environmental Conservation* 8(2):127–130.

Hubbell, S. P. (1979) "Tree Dispersion, Abundance, and Diversity in a Tropical Dry Forest." *Science* 203:1299–1309.

Independent Commission on International Development Issues. (1980) *North-South: A Programme for Survival*. London: Pan Books.

Intergovernmental Conference on Environmental Education, Tbilisi, USSR. (1977a) *Education and the Challenge of Environmental Problems*. UNESCO/ENVED, vol. 4. Paris: UNESCO-UNEP.

———. (1977b) *Final Report*. Paris: UNESCO-UNEP.

———. (1977c) *International Programme in Environmental Education*. UNESCO/ENVED, vol. 5. Paris: UNESCO-UNEP.

———. (1977d) *Needs and Priorities in Environmental Education: An International Survey*. UNESCO/ENVED, vol. 6. Paris: UNESCO-UNEP.

———. (1977e) *Regional Meetings of Experts on Environmental Education—A Synthetic Report*. UNESCO/ENVED, vol. 7. Paris: UNESCO-UNEP.

———. (1977f) *The United Nations Environmental Programme and Its Contribution to the Development of Environmental Education and Training*. UNEP/ENVED, vol. 9. Paris: UNESCO-UNEP.

International Foundation for Development Alternatives. (1979) *Third System Report: IFDA Dossier*. Nyon, Switzerland: IFDA.

International Union for Conservation of Nature and Natural Resources. (1972) *Red Data Book*, Vol. 1, *Mammalia*. Morges: IUCN.

———. (1975) *Red Data Book*, 5 vols. Gland, Switzerland: IUCN. (Looseleaf, issued periodically).

———. (1976) *An International Conference on Marine Parks and Reserves*. Morges: IUCN.

———. (1979) *Conservation for Thailand—Policy Guidelines*, 2 vols. [Proposal for National Environment Board]. Morges: Thailand in collaboration with UNEP.

International Union for Conservation of Nature and Natural Resources, with U.N. Environment Program and World Wildlife Fund. (1980) *World Conservation Strategy: Living Resources for Sustainable Development* Gland, Switzerland: IUCN/WWF; Nairobi: UNEP.

———. (1982) *United Nations List of National Parks and Protected Areas*. Gland, Switzerland: IUCN.

Janzen, D. H. (1966) "Coevolution of Mutualism between Ants and Acacias in Central America." *Evolution* 20:249–275.

———. (1969) "Allelopathy by Myrmecophytes:

The Ant *Azteca* as an Allelopathic Agent of Cecropia." *Ecology* 50(1):147–53.

Journal of International Affairs. (1977) Special Issue on the Global Commons. Vol. 31, No. 1 (Spring/Summer).

Khosla, A. (1980) "Proposal to Establish a New Development Promoting Non-Profit Corporation to Design, Produce, and Market Appropriate Technologies." Unpublished.

King, F. Wayne (1978a) Statement on the Endangered Species Authorization—H.R. 10883. In Endangered Species Oversight Hearings, U.S. House of Representatives, Subcommittee on Fish, Wildlife, Conservation, and the Environment. Endangered Species. Pt. 2, Ser. 95-40, pp. 807–811.

———. (1978b) "The Wildlife Trade." In *Wildlife and America*, ed. by H. P. Brokaw, Washington, D.C.: CEQ.

Krishna, R. (1980) "The Economic Development of India." *Scientific American* 243(3):134–143.

Leach, G. (1979) "Energy." Report prepared for the Agricultural Production: Research and Development in the 1980's Conference. Bonn: n.p.

Leontief, W. W. (1980) "The World Economy of the Year 2000." *Scientific American* 243(3):166–181.

Little, E. L. (1980) *The Audubon Society Field Guide to North American Trees: Eastern Region.* New York: Knopf.

Lovejoy, T. E. (1980a) "A Projection of Species Extinction." In CEQ and Department of State (1980), pp. 327–333.

———. (1980b) "Discontinuous Wilderness: Minimum Areas for Conservation." *Parks* 5(2):13–15.

———. (1982) "Designing Refugia for Tomorrow." In *Biological Diversification in the Tropics*, ed. by G. T. Prance, pp. 673–680. New York: Columbia University Press.

Lovejoy, T. E., and D. C. Oren. (1981) "The Minimum Critical Size of Ecosystems." In *Forest Island Dynamics in Man-Dominated Landscapes*, ed. by R. L. Burgess and D. M. Sharpe, pp. 7–12. New York: Springer-Verlag.

Lovejoy, T. E., and E. Salati. (1983) "Precipitating Change in Amazonia." In *The Dilemma of Amazonian Development*, ed. by E. Moran. Boulder: Westview Press.

Lovejoy, T. E., and H. O. R. Schubart. (1980) "The Ecology of Amazonian Development." In *Land, People, and Planning in Contemporary Amazonia*, ed. by F. Barbira-Scazzacchio, pp. 21–26. Cambridge: Cambridge University.

Lucas, G., and H. Synge. (1978) *The IUCN Plant Red Data Book.* Gland, Switzerland: IUCN.

Centro de Energia Nuclear na Agricultura (CENA) at Piracicaba, São Paulo, Brazil. (10–13 November 1981). Paper in preparation for *Acta Amazonica and the Environmentalist.*

Mar, B. W. (1980) "Dead Is Dead—An Alternative Strategy for Urban Water Management." Unpublished.

Marsh, George Perkins. (1864) *Man and Nature, or Physical Geography as Modified by Human Action.* New York: Scribner's.

Mathias, M. E. (1978) "The Importance of Diversity." Special Publication No. 1, vol. 11. Washington, D.C.: *American Association for the Advancement of Science.*

Mohl, B. (1980) Letter to World Wildlife Fund International, Denmark, 31 March.

Montgomery, G. E., and M. E. Sunquist. (1975) "Impacts of Sloths on Neotropical Forest Energy Flow and Nutrient Recycling." In *Tropical Ecological Systems*, ed. by F. B. Golley and E. Medina, pp. 67–98. New York: Springer-Verlag.

Morganthau, T., L. Donosley, W. J. Cook, J. Young, and D. Junkin. (1982) "The Disappearing Land." *Newsweek*, 23 Aug.

Myers, Norman. (1979) *The Sinking Ark.* New York: Pergamon Press.

———. (1981) "The Exhausted Earth." *Foreign Policy*, No. 42, Spring, pp. 1–6.

Nash, R. (1967) *Wilderness and the American Mind.* New Haven: Yale University Press.

National Academy of Sciences. (1975) *The Winged Bean: A High Protein Crop for the Tropics.* Washington, D.C.

———. (1977) *Guayule: An Alternative Source of Natural Rubber.* Washington, D.C.

———. (1980) *Firewood Crops: Shrub and Tree Species for Energy Production.* Washington, D.C.

Nature Conservation Council, Technical Sub-Committee. (1981) *Integrating Conservation and Development: A Proposal for a New Zealand Conservation Strategy.* Wellington, New Zealand.

Neary, J. (1981) "Pickleweed, Palmer's Grass, and Saltwort." *Science 81* 2(5):38–43.

Nye, P. H., and D. J. Greenland. (1965) *The Soil under Shifting Cultivation.* Farmham Royal: Commonwealth Agricultural Bureaux.

Ominde, S. N. (1972) "Migration and Child-bearing in Kenya." In *Population Growth and Economic Development in Africa*, ed. by S. N. Ominde and C. N. Ejiogu London: Heinemann.

Oppenshaw, K. (1980) "A Comparison of Metal and Clay Charcoal Cooking Stoves." Energy Report Series. Nairobi: UNEP.

Pimentel, D., and M. Pimentel. (1979) *Food, Energy and Society*. Kent: Edward Arnold.

Raven, P. H. (1981) Testimony presented to the U.S. Senate Committee on Environment and Public Works, Subcommittee on Environmental Pollution, Oversight Hearings on the Endangered Species Act of 1973. Mimeographed.

Reeves, Donald. (1980) "A Quaker Farmer Comments on Pope's Land Statement." *Catholic Rural Life* 29(3):21.

Roosevelt, Franklin D. (1937) Letter to all state governors regarding the Soil Conservation Service. 26 Feb.

Ruggieri, G. D. (1976) "Drugs from the Sea." *Science* 149:491–497.

Runte, A. (1979) *National Parks: The American Experience*. Lincoln: University of Nebraska Press.

SEBJ (1980) "Rapport d'etape du reseau de surveillance ecologique." *Direction Environnement*, p. 1.

Sagan, C., O. B. Toon, J. B. Pollack. (1979) "Anthropogenic Albedo Changes and the Earth's Climate." *Science* 206:1363–68.

Salati, E., A. Dall'Olio, E. Matsui, and J. R. Gat. (1979) "Recycling Water in the Amazon Basin: An Isotopic Study." *Water Resources Research* 15(5):1250–58.

Salati, E., P. Vose, and T. E. Lovejoy. (In press) "Precipitation and Water Recycling in Tropical Rain Forests with Special Reference to the Amazon basin." A position paper emanating from a workshop at the Centro de Energia Nuclear na Agricultura (CENA) at Piracicaba, São Paulo, Brazil. (10–13 November 1981). Paper in preparation for *Acta Amazonica and the Environmentalist*.

Salati, E., and E. Matsui. (1981) "Isotopic Hydrology in the Brazilian Amazon basin." In *Interamerican Symposium on Isotope Hydrology*, pp. 112–120. Bogotá: Instituto de Asuntos Nucleares.

Sampson, Neil R. (1981) *Farmland or Wasteland: A Time to Choose*. Emmaus, P.: Rodale Press.

Sassin, W. (1980) "Energy." *Scientific American* 243(3):106–117.

Sauvey, A. (1966) *General Theory of Population*. London: Weidenfeld and Nicolson.

Schubart, H. O. R. (1977) "Ecological Criteria for the Agricultural Development of the Dry Lands in Amazonia: Forest Ecology, Natural Resources including Wood Alcohol, Reclamation." *Acta Amazonica* 7(4):559–567.

Soule, M. E. (1980) "Thresholds for Survival: Maintaining Fitness and Evolutionary Potential." In Soule and Wilcox (1980),

Soule, M. E., and B. Wilcox, eds. (1980) *Conserva-tion Biology: An Evolutionary-Ecological Perspective*. Sunderland, Mass.: Sinauer Associates.

South Pacific Commission and South Pacific Bureau for Economic Cooperation. (1977) *Comprehensive Environmental Management Plan*. Noumea, New Caledonia: South Pacific Commission.

Stapp, William B. (1976) "International EE: The UNESCO-UNEP Programme." *Journal of Environmental Education* 8(2):19–25.

Suvanakorn, P. (1980) *The Administration of Forest Protection and Conservation in Thailand*. Bangkok: Royal Forest Department, Wildlife Conservation Division. Mimeographed.

Talbot, Lee M. (1957) "The Lions of Gir: Wildlife Management Problems of Asia." *Transactions of the North American Wildlife Conference* 22:570–579.

———. (1960) *A Look at Threatened Species: Conservation in the Middle East and Southern Asia*. London: Fauna Preservation Society.

———. (1964) "Wilderness overseas." In *Wildlands in Our Civilization*, San Francisco: Sierra Club.

Taylor, A. E., (trans.). (1929) *Plato, Timaeus and Critias*. London: Methuen.

Temple, S. A. (1977) "Plant-Animal Mutualism: Coevolution with Dodo Leads to Near Extinction of Plant." *Science* 197:885–886.

Thomas, William L., Jr., ed. (1956) *Man's Role in Changing the Face of the Earth*. Chicago: University of Chicago Press.

Thornback, J., and M. Jenkins. (1982) Red Data Book, *Mammals*. Part 1: Threatened Mammalian Taxa of the Americas and the Australasian Zoogeographic Region (excluding Cetacea). Gland, Switzerland: IUCN.

Udvardy, Miklos. (1975) *A Classification of the Biogeographical Provinces of the World*. IUCN Occasional Paper 18. Morges: IUCN.

United Nations. (1980) *Selected Demographic Indicators by Country, 1950–2000: Demographic Estimates and Projections as Assessed in 1978*. New York: United Nations.

———. (1981a) *Commodity Trade Statistics*. New York: United Nations, Department of International Economic and Social Affairs.

———. (1981b) *Statistical Yearbook 1979/1980*. New York: United Nations Department of International Economic and Social Affairs.

U.N. Educational, Scientific, and Cultural Organization. (1968) *Use and Conservation of the Biosphere*, vol. 10. Paris: UNESCO, Natural Resources Research.

———. (1971) *International Coordinating Council of the Programme on Man and the Biosphere*. (MAB Final Report). Paris.

————. (1972) "Protection of the World Cultural and Natural Heritage." *UNESCO Resolutions of 1972.* Paris.

U.N. Food and Agriculture Organization. (1977) *The Fourth World Food Survey.* Rome: FAO.

U.N. General Assembly. (1972) *Report of the United Nations Conference on the Human Environment.* A/Conf. 48/14. New York.

————. (1979) Resolution A/Res/34/188. New York.

U.S. Congress. (1935) Soil Conservation Act. Public Law 74-76. *University of Miami Law Review.* (1978) Special Issue on Antarctic Resources: A New International Challenge, Vol. 33, No. 2.

van Buren, A. E. (1980) "The Chinese Development of Biogas and Its Applicability to East Africa." Energy Report Series. Nairobi: UNEP.

Vietmeyer, N. (1978) "Revegetation Using Selected Species." In *Proceedings of the U.S. Strategy Conference on Tropical Deforestation*, ed. by U.S. Department of State and U.S. Agency for International Development, pp. 48–50. Washington, D.C.

Waldheim, Kurt. (1980) Message on the occasion of the launching of the World Conservation Strategy on 5 March 1980. New York: United Nations, Office of the Secretary General.

Walker, E. P. (1975) *Mammals of the World*, 3rd ed., Vol. 2. Baltimore: Johns Hopkins University Press.

Western D. (1976) "A New Approach to Amboseli." *Parks* 1(2):1–4.

————. (1982) "The Environment and Ecology of Pastoralists in Arid Savannahs." *Development and Change*, April.

Western, D., and W. Henry. (1979) "Economics and Conservation in Third World National Parks." *Bioscience* 29(7):414–418.

Western, D., and J. Ssemakula. (1980) "The Present and Future Patterns of Consumption and Production of Wood Energy in Kenya." *Energy Report Series.* Nairobi: UNEP.

Wijewardene, R. (1978) "Appropriate Technology in Tropical Farming Systems." *World Crops*, May/June.

Wilson, E. O. (1980) "Resolutions for the Eighties." *Harvard Magazine*, Jan.–Feb., p. 21.

Winder, D. (1982) "Why the World Population Explosion Fizzled: A Tale of Sharply Changed Village Attitudes." *Christian Science Monitor*, 14 Sept., pp. 1, 7.

World Bank. (1982) *World Development Report 1982.* New York: Oxford University Press.

Contributors

The Editors

Francis R. Thibodeau *is Director of Science at the Center for Plant Conservation, the United States repository for endangered plant species. He also holds appointments at Tufts University in the Department of Urban and Environmental Policy and at the Arnold Arboretum of Harvard University. He is a specialist in both plant ecology and the analysis of environmental policy.*

Hermann H. Field *is professor emeritus in environmental planning at Tufts University and the founder of its Department of Urban and Environmental Policy. He organized the 1981 Tufts course on the World Conservation Strategy and was co-director in 1983 of the first Talloires seminar: Environmental Planning in the Context of Development Investment. He is a member of IUCN's Commission on Environmental Planning and is a fellow of the American Institute of Architects.*

The Authors

Harold J. Coolidge *is honorary president of IUCN and honorary chair of the Coolidge Center for Environmental Leadership. A director emeritus of World Wildlife Fund—US, he has had a lifetime interest in international conservation and primatology. He is an honorary member of IUCN's Commission on National Parks and Protected Areas and of the Species Survival Commission.*

Lee M. Talbot *is a fellow of the East-West Environment and Policy Institute and visiting fellow at the World Resources Institute. Previously he was director general of IUCN, and before that a member of its executive board and vice president. With Sidney Holt he co-authored* New Principles for Conservation of Wild Living Resources.

Raymond F. Dasmann *is professor of environmental studies at the University of California at Santa Cruz. He is a former director of international programs at the Conservation Foundation and a former senior ecologist at IUCN. He is the author of* Environmental Conservation *and of* Wildlife Biology. *He is a member of IUCN's Commission on Environmental Planning.*

Gerard A. Bertrand *is president of the Massachusetts Audubon Society and former chief of international affairs of the U.S. Fish and Wildlife Service. He is a former chair and present member of the executive committee of the American Committee for International Conservation. He is a member of IUCN's Species Survival Commission.*

F. Wayne King *is director of the Florida State Museum and a former director of zoology and conservation of the New York Zoological Society. His special fields are conservation of biological diversity and the systemics, ecology, and behavior of reptiles. He is deputy chair of IUCN's Species Survival Commission.*

Gerardo Budowski *is head of the natural renewable resources program at the Tropical Agricultural Center for Research and Training (CATIE) in Costa Rica. He is a former director general of IUCN and prior to that was active in the Man and the Biosphere Program at UNESCO. He has published widely in his special field of tropical forest ecology, land use, and conservation for development. He is a member of IUCN's Commission on Environmental Policy, Law, and Administration.*

Kenton R. Miller *is director general of IUCN. He is a specialist in wildland management and biosphere reserves, with wide experience in the Caribbean and Latin America. He is presently on leave from his former position as associate professor of natural resources at the University of Michigan. He is a member of IUCN's Commission on National Parks and Protected Areas.*

Peter Jacobs *is associate dean of the Faculté de l'Amenagement and professor of landscape architecture and regional planning at the University of Montreal. A member of numerous expert committees and commissions, he is chair of public review of Canada's Lancaster Sound Regional Study. He is author of* Environmental Strategy and Action *and is chair of IUCN's Commission on Environmental Planning.*

William Stapp *is professor of natural resources and chair of the behavior and environment program at the* School of Natural Resources, University of Michigan. *His special field is international environmental education. A former vice president of the U.N. World Conference on Environmental Education, he is a member of the steering committee of IUCN's Commission on Education.*

David Western *is resource ecologist for the New York Zoological Society and special ecological adviser to the African Wildlife Leadership Foundation. He has an intimate knowledge of rural Africa and has authored many research reports and articles on the ecology of African ecosystems, wildlife biology, and animal populations.*

Wolfgang Burhenne *is secretary general of the Interparliamentary Working Center in Bonn. He is editor of* Environmental Policy and Law *of the International Council of Environmental Law and co-author with Will A. Irwin of* World Charter for Nature—A Background Paper. *He is chair of IUCN's Commission on Environmental Policy, Law, and Administration.*

Malcolm Forster *is professor of law at the University of Southampton, England, and director of its Center for Energy, Law, and Policy. He is counsel for the Seychelles to the International Whaling Commission and a member of IUCN's Commission on Environmental Policy, Law, and Administration.*

Alexandre Kiss *is director of research at the Centre National de la Recherche Scientifique and president of the European Council for Environmental Law. He is the author of* Survey of Current Development in International Environmental Law. *He is a member of IUCN's Commission on Environmental Policy, Law, and Administration.*

Ann L. Hollick *is currently a National Science Foundation visiting professor at MIT's Center for International Studies. She has held positions with the U.S. Department of State's Economic and Commercial Bureau and its Oceans, Environmental, and Scientific Affairs Bureau. She is author of* U.S. Foreign Policy on the Law of the Sea.

Thomas E. Lovejoy *is vice president for science at* World Wildlife Fund—US *and chair of the Wildlife Pres-*

ervation Trust International. A biologist by profession, he is a member of the Oversight Committee of the Rockefeller Brothers Fund/Environmental Defense Fund project on tropical forests and species resources. He has a book in preparation on people and the biosphere. He is chair of the Tropical Forest Working Group of IUCN's Commission on Ecology.

Alexander R. Brash *was a research assistant to Dr. Lovejoy and is currently in the graduate program at the Yale School of Forestry and Environmental Studies.*

Sidney Holt *has been Senior Overseas Scholar at St. John's College, Cambridge University and visiting professor of environmental studies at the University of California at Santa Cruz. He is executive director of the International League for the Protection of Cetaceans, and former director of the FAO's Division of Fisheries Resources as well as secretary of UNESCO's Intergovernmental Oceanographic Commission. In IUCN he is chair of its Committee on Marine Mammals and member of its Species Survival Commission.*

Jean Mayer *is president of Tufts University. From 1950 to 1976 he was professor of nutrition at Harvard University. He is a former consultant to UNICEF, FAO, and WHO. In 1969 he was chair of the White House Conference on Food, Nutrition, and Health. From 1978 to 1980 he was vice chair and acting chair of the Presidential Commission on World Hunger.*

Jean-Claude Faby *is deputy director of the United Nations Environment Program's New York Liaison Office. Previously he was information officer of the U.N. Office of Public Information.*

In addition to the principal contributors, the following panelists contributed substantially to the discussion following each chapter.

An Introduction to World Conservation

Deborah V. Howard,
Director of Environmental Affairs of the Massachusetts Audubon Society

Richard Kendall,
Commissioner of the Massachusetts Department of Environmental Management

Norton H. Nickerson,
Professor of Biology at Tufts University

Ecological Processes and Life Support Systems

Ernest M. Gould,
Forester at the Harvard Forest and Professor of Biology at Harvard University

Dean Johnson,
Director of the Boston Harbor Associates

Robert Yaro,
Deputy Commissioner of the Massachusetts Department of Environmental Management

Preservation of Genetic Diversity

Peter S. Ashton,
Director of the Arnold Arboretum and Professor of Dendrology at Harvard University

Bradford Northrop,
Eastern Regional Director of the Nature Conservatory

William O. Satterfield,
Resident Veterinarian at the Franklin Park Zoo of the Boston Zoological Society

Sustainable Use of Species and Ecosystems

Robert Howarth,
Staff Scientist at the Ecosystems Center of the Woods Hole Marine Biological Laboratory

Deborah V. Howard,
Director of Environmental Affairs of the Massachusetts Audubon Society

Paul Nickerson,
Senior Staff Specialist, New England Regional Office of the U.S. Fish and Wildlife Service

National and Regional Conservation Strategies

Charles W. Harris,
Professor of Landscape Architecture at Harvard University

Frank Schnidman,
Visiting Scholar at the Harvard Law School

Environmental Planning and Rational Use

Kelly McKlintock,
Executive Director of the Massachusetts Forest and Park Association

Alfred Rubin,
Professor of International Law at the Fletcher School of Law and Diplomacy of Tufts University

C. Dart Thalman,
Coordinator of Interregional Exchange and Policy at the Atlantic Center for the Environment

Building Support for Environmental Education

Nancy Anderson,
Director of Environmental Programs for the Lincoln Filene Center of Tufts University

N. Bruce Hanes,
Professor of Civil Engineering at Tufts University

Lawrence Susskind,
Chairman of the Department of Urban Studies and Planning at the Massachusetts Institute of Technology

Naseeb Dajani,
Executive Officer of the IUCN Education Commission

Conservation-Based Rural Development

Albert Baez,
Chairman of the IUCN Education Commission

Helen L. Vukasin,
Coordinator of the Environment and Development Program, CODEL Inc.

Malcolm Odell,
Development Consultant for Synergy International

Environmental Policy and Law

Thomas Arnold,
Director of the New England Rivers Center

Gregor McGregor,
Vice-President of the Massachusetts Association of Conservation Commissions

Stephen M. Leonard,
Chief of the Environmental Protection Division of the Massachusetts Department of the Attorney General

Management of the Global Commons

William H. Bossert,
Gordon McKay Professor of Applied Mathematics at Harvard University

Newall B. Mack,
Researcher at the Energy and Environmental Policy Center of Harvard University

Malcolm Tink Taylor,
President of the Squam Lake Association

Tropical Forests and Genetic Resource Areas

Ernest M. Gould,
Forester at the Harvard Forest, and Professor of Biology at Harvard University

Stuart B. Avery,
Chairman of the Wildlife Committee, National Sierra Club

Lynn Margulis,
Professor of Biology at Boston University

John Todd,
Director of the New Alchemy Institute

Regional Strategies for
Managing the Oceans

Thomas Arnold,
Director of the New England Rivers Center

John Ehrenfeld,
Director of the New England River Basins Commission

John H. Prescott,
Executive Director of the New England Aquarium

Phoebe Wray,
Executive Director of the Center for Action on Endangered Species

Index